T0282631

CAMBRIDGE LIBRARY COLLECTION

Books of enduring scholarly value

Earth Sciences

In the nineteenth century, geology emerged as a distinct academic discipline. It pointed the way towards the theory of evolution, as scientists including Gideon Mantell, Adam Sedgwick, Charles Lyell and Roderick Murchison began to use the evidence of minerals, rock formations and fossils to demonstrate that the earth was older by millions of years than the conventional, Bible-based wisdom had supposed. They argued convincingly that the climate, flora and fauna of the distant past could be deduced from geological evidence. Volcanic activity, the formation of mountains, and the action of glaciers and rivers, tides and ocean currents also became better understood. This series includes landmark publications by pioneers of the modern earth sciences, who advanced the scientific understanding of our planet and the processes by which it is constantly re-shaped.

Volcanoes and Earthquakes

The original French edition of this book appeared in 1866 as part of Hachette's extensive, popularising Bibliothèque des Merveilles series, which included several science titles by Frédéric Zurcher (1816–90) and Elie Margollé (1816–84). Their books were illustrated with attractive wood-cuts, and remained in print until the 1880s; they were also translated into English. This volume was published in London in 1868, and is a good example of popular science publishing in Victorian Britain. The material is organised geographically, beginning in Europe with Vesuvius, Etna and Icelandic volcanoes including Hecla, all of which had recently seen major eruptions. The authors quote from eyewitness accounts, and refer to scholarly publications on volcanoes including Darwin (1844) and Scrope (1862), also reissued in the Cambridge Library Collection. Later chapters describe oceanic volcanoes, the Andes, the volcanic zone of New Zealand's North Island, and recently discovered volcanoes such as Mt Erebus in Antarctica.

Cambridge University Press has long been a pioneer in the reissuing of out-of-print titles from its own backlist, producing digital reprints of books that are still sought after by scholars and students but could not be reprinted economically using traditional technology. The Cambridge Library Collection extends this activity to a wider range of books which are still of importance to researchers and professionals, either for the source material they contain, or as landmarks in the history of their academic discipline.

Drawing from the world-renowned collections in the Cambridge University Library and other partner libraries, and guided by the advice of experts in each subject area, Cambridge University Press is using state-of-the-art scanning machines in its own Printing House to capture the content of each book selected for inclusion. The files are processed to give a consistently clear, crisp image, and the books finished to the high quality standard for which the Press is recognised around the world. The latest print-on-demand technology ensures that the books will remain available indefinitely, and that orders for single or multiple copies can quickly be supplied.

The Cambridge Library Collection brings back to life books of enduring scholarly value (including out-of-copyright works originally issued by other publishers) across a wide range of disciplines in the humanities and social sciences and in science and technology.

Volcanoes and Earthquakes

Frédéric Zurcher
Elie Margollé
Translated by
Winifred Lockyer

CAMBRIDGE UNIVERSITY PRESS

Cambridge, New York, Melbourne, Madrid, Cape Town,
Singapore, São Paolo, Delhi, Mexico City

Published in the United States of America by Cambridge University Press, New York

www.cambridge.org
Information on this title: www.cambridge.org/9781108049405

This edition first published 1868
This digitally printed version 2012

ISBN 978-1-108-04940-5 Paperback

CONTENTS.

V

THE ANDES.

VI

VOLCANOES OF THE PACIFIC AND INDIAN OCEANS.

VII

PREHISTORIC AND LUNAR VOLCANOES.

VIII

EARTHQUAKES.

IX

MUD VOLCANOES—SPRINGS AND WELLS OF FIRE—THERMAL SPRINGS.

X

UPHEAVALS.

LIST OF ILLUSTRATIONS.

VOLCANOES AND EARTHQUAKES.

I

VESUVIUS.

FIRST ERUPTION—DEATH OF PLINY—HERCULANEUM AND POMPEII—ERUPTIONS OF 1631, 1737, 1822 AND 1858—ASCENTS OF THE MOUNTAIN.

FIRST ERUPTION.

THE ROMANS believed that in former times Vesuvius had been in a state of eruption, but the accounts handed down from very distant periods were nearly forgotten, and people inhabited the towns built on its slopes without fear. 'Above those places,' says Strabo,* speaking of Herculaneum and Pompeii, 'lies Mount Vesuvius, encompassed by the finest-cultivated fields except on its summit. This is, indeed, mostly a plain surface, but unfruitful, and has an ashy appearance. It shows cleft hollows of sooty-looking rock, appearing as if it had been eaten into by the action of fire; so that we may conjecture that, at some former time, there burnt in these orifices a

* Strabo, lib. v. p. 247, Casaub.; quoted in Humboldt's *Cosmos*, vol. iv. p. 407, Sabine's translation.

fire which has now become extinct, the fuel which supported it having been consumed.'

The servile war which broke out in the Campagna, in the year 73 before our era, and which held the consular armies so long in check, commenced by the revolt of two hundred gladiators, Gauls and Thracians, Spartacus being their chief. Having taken refuge on Vesuvius, they were there attacked by troops sent from Rome; but they remained in safety in one of the ravines of the mountain, from which they were able to take in rear the quarters of the besiegers, who, seeing themselves surrounded, took to flight, and left their camp in the power of the enemy.

The volcano, in spite of its long repose, was not extinguished. It broke out all at once in a formidable eruption, which buried several towns lying at its foot. This happened in the month of August 79, after violent earthquakes, which, during the preceding sixteen years, had devastated the country. Pliny the younger, in the following letter, addressed to the historian Tacitus, recounted this event, in the midst of which his uncle perished, a victim to his humanity and his love of science.

*Pliny the Younger's account of the Death of the Elder Pliny.**

'Your request that I would send you an account of my uncle's death, in order to transmit a more exact relation of it to posterity, deserves my acknowledg-

* *Pliny's Letters*, vol. i. book vi. p. 325, Melmoth's translation.

ments; for if this accident shall be celebrated by your pen, the glory of it, I am well assured, will be rendered for ever illustrious. And notwithstanding he perished by a misfortune, which, as it involved at the same time a most beautiful country in ruins, and destroyed so many populous cities, seems to promise him an everlasting remembrance; notwithstanding he has himself composed many and lasting works; yet I am persuaded, the mentioning of him in your immortal writings will greatly contribute to eternise his name. Happy I esteem those to be whom providence has distinguished with the abilities either of doing such actions as are worthy of being related, or of relating them in a manner worthy of being read; but doubly happy are they who are blessed with both these uncommon talents; in the number of which my uncle, as his own writings and your history will evidently prove, may justly be ranked. It is with extreme willingness, therefore, I execute your commands; and should, indeed, have claimed the task if you had not enjoined it.

'He was at that time with the fleet under his command at Misenum.* On August 24, about one in the afternoon, my mother desired him to observe a cloud which appeared of a very unusual size and shape. He had just returned from taking the benefit of the sun, and after bathing himself in cold water, and taking a slight repast, was retired to his study; he immediately arose, and went out upon an eminence, from whence he might more distinctly

* In the Gulf of Naples.

view this very uncommon appearance. It was not at that distance discernible from what mountain this cloud issued, but it was found afterwards to ascend from Mount Vesuvius. I cannot give you a more exact description of its figure than by resembling it to that of a pine-tree, for it shot up a great height in the form of a trunk, which extended itself at the top into sort of branches; occasioned, I imagine, either by a sudden gust of air that impelled it, the force of which decreased as it advanced upwards, or the cloud itself being pressed back again by its own weight, expanded in this manner; it appeared sometimes bright and sometimes dark and spotted, as it was either more or less impregnated with earth and cinders. This extraordinary phenomenon excited my uncle's philosophical curiosity to take a nearer view of it. He ordered a light vessel to be got ready, and gave me the liberty, if I thought proper, to attend him. I rather chose to continue my studies; for, as it happened, he had given me an employment of that kind. As he was coming out of the house,* he received a note from Rectina, the wife of Bassus, who was in the utmost alarm at the imminent danger which threatened her; for her villa being situated at the foot of Mount Vesuvius, there was no way to escape but by sea; she earnestly intreated him, therefore, to come to her assistance. He accordingly changed his first design, and what he began with a philosophical, he pursued with an heroical turn of

* The manuscript and printed copies vary extremely from each other as to the reading of this passage.

mind. He ordered the galleys to put to sea, and went himself on board with the intention of assisting not only Rectina, but several others; for the villas stand extremely thick upon that beautiful coast. When hastening to the place from whence others fled with the utmost terror, he steered his direct course to the point of danger, and with so much calmness and presence of mind, as to be able to make and dictate his observations upon the motion and figure of that dreadful scene.

'He was now so nigh the mountain that the cinders, which grew thicker and hotter the nearer he approached, fell into the ships, together with pumice-stones and black pieces of burning rock. They were likewise in danger not only of being aground by the sudden retreat of the sea, but also from the vast fragments which rolled down from the mountain and obstructed all the shore. Here he stopped to consider whether he should return back again; to which the pilot advising him, "Fortune," said he, "befriends the brave. Carry me to Pomponianus!" Pomponianus was then at Stabiæ,* separated by a gulf, which the sea, after several insensible windings, forms upon that shore. He had already sent his baggage on board; for, though he was not at that time in actual danger, yet being within the view of it—and, indeed, extremely near if it should in the least increase—he was determined to put to sea as soon as the wind should change. It was favourable, however, for carrying my uncle to Pomponianus,

* Now called Castel a Mar di Stabia, in the Gulf of Naples.

whom he found in the greatest consternation. He embraced him with tenderness, encouraging and exhorting him to keep up his spirits; and the more to dissipate his fears he ordered, with an air of unconcern, the baths to be got ready. When, after having bathed, he sat down to supper with great cheerfulness, or at least (what is equally heroic) with all the appearance of it.

'In the meanwhile, the eruption from Mount Vesuvius flamed out in several places with much violence, which the darkness of the night contributed to render still more visible and dreadful. But my uncle, in order to soothe the apprehensions of his friend, assured him it was only the burning of the villages, which the country people had abandoned to the flames. After this he retired to rest, and it is most certain he was so little discomposed as to fall into a deep sleep; for being pretty fat, and breathing hard, those who attended without actually heard him snore. The court which led to his apartment being now almost filled with stones and ashes, if he had continued there any time longer, it would have been impossible for him to have made his way out; it was thought proper, therefore, to awaken him. He got up, and went to Pomponianus and the rest of his company, who were unconcerned enough to think of going to bed. They consulted together whether it would be most prudent to trust to the houses, which now shook from side to side with frequent and violent concussions; or fly to the open fields, where the calcined stones and cinders, though light indeed, yet fell in large showers, and threatened destruction. In this distress they resolved for the fields, as the less dangerous situation of the two—a resolution which, while the rest of the company were hurried into by their fears, my uncle embraced upon cool and deliberate consideration. They went out then, having pillows tied upon their heads with napkins, and this was their whole defence against the storm of stones that fell round them. It was now day everywhere else, but there a deeper darkness prevailed than in the most obscure night, which, however, was in some degree dissipated by torches and other lights of various kinds. They thought proper to go down farther upon the shore, to observe if they might safely put out to sea, but they found the waves still ran extremely high and boisterous. There my uncle, having drunk a draught or two of cold water, threw himself down upon a cloth which was spread for him, when immediately the flames, and a strong smell of sulphur, which was the forerunner of them, dispersed the rest of the company, and obliged him to arise. He raised himself up with the assistance of two of his servants, and instantly fell down dead—suffocated, as I conjecture, by some gross and noxious vapour, having always had weak lungs, and frequently subject to a difficulty of breathing. As soon as it was light again, which was not till the third day after this melancholy accident, his body was found entire, and without any marks of violence upon it, exactly in the same posture that he fell, and looking more like a man asleep than dead.'

HERCULANEUM AND POMPEII.

The fall of pumice-stones at the beginning of the eruption shows that the immense cloud projected by the gas of the new crater was formed both of the ashes from the depths of the earth and of the *débris* of a great portion of the old cone of Vesuvius, to which they give the name of Somma. The destruction of the towns of Herculaneum, Pompeii, and Stabiæ is generally accounted for by a long-continued fall of these materials; but the transport of beds of such thickness is difficult to imagine, by reason of the distance which separates them from the crater, and the idea propounded on this subject by M. Ch. Sainte-Claire Deville appears to us much more probable. This learned explorer of volcanoes shows us, indeed, that at the moment when Vesuvius again became active, the crest split, following transverse fissures, which he has shown to be connected with the volcanic system of the Campana, and that two amongst them passed exactly by the destroyed towns, which, according to this theory, would have been buried with ashes, mud, and lava ejected from these orifices. It is known that, until the middle of the last century, the real sites of these towns were not discovered. By a series of excavations since that period, the people of modern times have been transported, as if by magic, into the middle of ancient life, and have been enabled to trace most precious revelations, both to science and history, in the ruins preserved for eighteen centuries by the volcanic strata.

A very interesting book by M. Marc Monnier* gives a description of these ruins. Monuments, edifices, and a thousand objects of art or industry, have been disinterred. During the last few years human forms have been discovered; but, such sad forms! The ashes, softened by the steam of the eruption, moulded themselves round the bodies at the very instant they expired. A very simple process has reproduced these once human forms in plaster.

'Nothing is more striking,' remarks M. Marc Monnier, 'than this spectacle. They are not statues, but human bodies moulded by Vesuvius; the skeletons are still there, in these envelopes of plaster, which reproduce that which time would have destroyed—that which the damp ashes have preserved—the garments and the flesh, one could almost say the life. The bones here and there pierce certain places where the stream could not reach. There exists nowhere anything similar to this. The Egyptian mummies are naked, black, hideous; they were arranged for an eternal repose in a consecrated attitude. But the disinterred Pompeians are human beings whom one sees die.'

ERUPTIONS OF 1631, 1737, 1822 AND 1858.

There exist indications of eruptions since 79 in the years 204, 472, 512, 685, 993, 1036, 1136. That of 1136 was very violent, but the volcano after that remained inactive for nearly five hundred years. At

* *Pompéi et les Pompéiens.*

the beginning of the seventeenth century the summit had the form of a large basin, which, according to the testimony of travellers, was covered with old oaks, chestnuts, and maple-trees.

During the month of December 1631, the volcano opened underneath the vast depression which separates the crater from the Somma, and which is called l'Atrio del Cavallo. A great portion of the mountain fell in, and the eruption ended with a stream of lava which disappeared in the sea, near to Portici, after having burnt houses and trees in its path. The volcano broke out again in 1660, and underwent great changes of form until 1685. The years 1707 and 1724 also indicate periods of activity.

In May 1737, the mountain sent forth a great quantity of smoke, and from the 16th to the 19th of that month subterraneous rumblings were heard, accompanied by noisy detonations.

'On Monday the 20th, at the thirteenth hour, the mountain made so loud an explosion that the shock was strongly felt, not only in the neighbourhood but also in the cities twelve miles round. Black smoke, intermixed with ashes, was seen suddenly to rise in vast curling globes, which spread wider as it moved farther from the basin. The explosions continued very loud and frequent all this day, shooting up very large stones through the thick smoke and ashes, about a mile high, to the horror of the beholders and danger of all the neighbouring buildings.

'At the twenty-fourth hour of the same day, Monday, May 20, amidst the noise and dreadful shocks,

Fig. 2.—ERUPTION OF VESUVIUS IN 1737

the mountain burst on the first plain, a mile distant obliquely from the summit, and there issued from the new opening a vast torrent of fire; whence, by the quantity of fire incessantly thrown up into the air at a distance, all the south side of the mountain seemed in a flame. The liquid torrent flowed out of the new vent, rolling along the plain underneath, which is about a mile long and nearly four miles broad, and in its way it spread very speedily near a mile wider, and by the fourth hour of the night it reached the end of the plain and to the foot of the low hills situate to the south; but as these hills are rugged with rocks, the greatest part of the torrent ran down the declivities between these rocks and into two valleys, falling successively into the other plain which forms the basis of the mountain, and, after uniting there, it divided into four lesser torrents, one of which stopped in the middle of the road a mile and a half distant from the Torre del Greco; the second flowed into a large valley; the third ended under the Torre del Greco, near the sea; and the fourth at a short distance from the new mouth.

'The torrent which flowed into the valley ran as far as between the Church of the Carmelites and that of the Souls of Purgatory, by the eighth hour on Tuesday. The matter of the torrent ran like melted lead; in eight hours it made four miles, and consequently it flowed half a mile in an hour—a new and remarkable circumstance of this eruption, seeing Bulifone thought it very strange; but in the eruption of 1698 the torrent advanced sixty paces in an hour, whence he infers that such great swiftness proceeded from a greater degree of liquidation of the matter. The trees which the torrent lighted on in its way, upon the first touch took fire, and fell under the weight of the matter. . . . Sixteen days afterwards the matter continued hot, and was very hard, but it was broken by repeated blows.

'A piece of glass fastened on the top of a pole, and thrust into this matter, was in four minutes reduced to a paste. Under the mass of the torrent were heard frequent reports, which made the church shake as if by an earthquake. Along the whole surface of the torrent there appeared small fissures, out of which issued smoke that stunk of brimstone mixed with sea-water; yet these exhalations are not poisonous, but rather a remedy for some diseases. The stones round about these fissures were observed to be covered with sublimed salts, the nature of which I shall explain hereafter. Iron thrust into these fissures, was taken out moist; but upon thrusting in paper, it was not moistened but rather somewhat hardened.

'At the same time when the new mouth opened, that on the summit of the mountain vomited a vast quantity of burning matter, which, dividing into torrents and small streams, ran partly towards the Salvadore, and partly towards Ottajano; and at the same time that this matter issued out, red-hot stones were seen to be cast out of the mouth, in the midst of black smoke, frequent flashes of lightning, and thunder, all produced by the same matter. These impetuous expulsions of fire continued till Tuesday, when the eruptions of the melted matter, the flashes, and thundering noise ceased; but a strong south-west wind arising, the ashes were carried in great quantities to the utmost boundaries of the kingdom—in some places very fine, in others as coarse as Ischian sand; and in the neighbourhood they not only felt this plentiful shower of ashes, but likewise pieces of pumice-stone, and other large stones.

'On Tuesday night the fury of the mountain began to abate, so that on Sunday there was scarcely any flame seen to break out of the upper mouth, and on Monday but little smoke and ashes. This day it began to rain plentifully, which continued to Tuesday, and afterwards for many days: a circumstance which has constantly happened after eruptions of times past.

'The damages done in the neighbourhood by this eruption of fire and ashes are incredible. At Ottajano, situated between four and a half and five miles from Vesuvius, the ashes on the ground were four palms high. All the trees were burnt (or blasted), the people terribly affrighted, and many houses crushed by the weight of the ashes and stones that fell.'

It was near Torre del Greco that in 1797 a river of lava 1,500 feet wide and 14 feet high, flowed three miles and a half, then extended 600 feet into the sea. The English ambassador, Sir William Hamilton, who has left interesting accounts of Vesuvius, took a boat and was conducted near this burning wall. 'At a distance of 300 feet,' said he, 'the lava caused the water to smoke and boil: the fish perished as far as two miles beyond.'*

In 1822, the eruption was preceded by a sinking down of the summit. The cone, which rose to a height of 218 yards above the floor of the crater, and which was visible above its wall, gave way in the night of October 22, with a horrible noise. 'The following night,' says Humboldt,† 'a fiery eruption of ashes and rapilli‡ commenced. It lasted twelve days without interruption; the first four days it attained its maximum. During this time, the detonations in the interior of the volcano were so violent, that the simple disturbance of the air (no shock of the earth had been felt) cracked the ceilings in the apartments of the Portici palace. The neighbouring villages, Resina, Torre del Greco, Torre dell' Annunziata, and Bosco Tre Case, were the scene of a curious spectacle: the atmosphere was so filled with ashes that all the country in midday, during several hours, was enveloped in the most profound darkness. The people walked about the streets with lanterns, as happens often at Quito during the eruptions of Pichincha. Among the inhabitants there was such a *sauve qui peut* as had never before occurred. They feared less the torrents of lava than an eruption of ashes; this eruption, violent and unusual, added to the vague tradition of the fate of Herculaneum, Pompeii, and Stabiæ, terrified the imagination.

* *Philosophical Transactions.* † *Aspects of Nature.*

‡ Fragments of incandescent pumice-stones.

'The steam from the hot water which rose from the crater condensed at the contact of the atmosphere into a thick cloud 9,000 feet high. This instant condensation of the vapour, and the formation also of the cloud, increased the electric tension. Lightning traversed the column of ashes in every direction, and the rolling of thunder was distinctly heard, in addition to the internal noise of the volcano. In no other eruptions has electricity been manifested in so striking a manner. The morning of October 26, a strange rumour was spread abroad: it was that a torrent of boiling water was ejected from the crater, and was precipitated down the cone of ashes. However Monticelli, the learned and zealous observer of the volcano, soon observed that this rumour was the result of an illusion. The supposed torrent was only an immense quantity of dry ashes which came forth like moving sand from a crack in the upper edge of the crater. The eruption of Vesuvius was preceded by a drought which desolated the fields; it ended by the volcanic storm before mentioned, followed with a long heavy rain. In all parts this characterises the end of an eruption.' *

In 1850, the lava came forth from the crater in extraordinary abundance, bringing with it large blocks of granite. The edge of the vast plateau formed by this stream became a kind of cyclopean rampart, raised more than five miles above the plain where the torrent stopped.

From 1855 until 1858 Vesuvius was continually in

* Humboldt. *Personal Narrative.*

eruption. At the end of May and the beginning of June in the latter year, the eruption continued with great violence. In the space of two days five fissures, vomiting an enormous quantity of lava and smoke, opened on the sides of the cone. The following view, from a drawing made at that time, represents several of these fissures, situated near the

Fig. 3.—Eruption of Vesuvius in 1858.

base, at the time of their greatest activity. It was a magnificent spectacle in the middle of the night. The lava formed large rivers, which divided again into many branches. M. Palmieri, director of the Observatory built on the mountain, has carefully described all the phenomena of this eruption, also those which accompanied the violent eruption of Torre del Greco in 1861.

ASCENTS OF THE MOUNTAIN.

We have had occasion to visit Vesuvius twice. First in 1836, during a period of tranquillity. After leaving the village of Resina on the sea-coast, we passed through the vineyards which produce the celebrated wine Lacryma-Christi. Ancient streams of lava were met with at intervals, but partly clothed with fresh verdure. A group of large trees still shades the plateau of the hermitage, situated half-way up the mountain. The vegetation afterwards diminishes, and we enter a region purely mineral. The immense desert of scoriæ and ashes extends on every side.

The aspect of these lava fields is indeed especially striking. As in the Alpine glaciers, it has the aspect of a sea, with its waves suddenly congealed. But instead of a crystal shining under the sun's rays, we have before us a black dull matter, on which lie, here and there, grey or yellowish scoriæ like crests of foam.

It is necessary to pass some time in the valley which is formed by the Somma and the cone, in order to find the point from which side the ascent of the latter is easiest. On account of the great inclination of the slope, it is always difficult, and takes nearly an hour. In the lower part scoriæ predominate. The *débris* on which we walk often gives way, and obliges us to go over the ground again. Higher up, the feet sink into a fine ash, which renders the progress extremely wearisome. The summit reached, we command a view of the crater, and the eyes are carried alternately from the horribly shattered abyss to the harmonious landscape; the Gulf of Naples, with its isles and promontories. In the words of a poet

It is Paradise viewed from Hell.

At each eruption, the large basin which forms the crater changes form. Fresh cones are thrown up,

Fig. 4.—Vesuvius (1866).

and again fall away, rocks are superposed in falling, and crevasses opened. We were able to observe the bottom of some of them, in which the matter presented remarkable varieties of colour; and the mouth even of the volcano, from which issued a thick column of smoke. Circumstances were favourable to us. The wind cleared the side of the abyss on which we ascended, and facilitated the examination of the interior walls, covered with calcined stones and vitrifactions. A few minutes sufficed to reach the base of the cone, mounted with so much trouble. We slid on the ashes as on the snows of high mountains.

The return to the midst of living nature had a great charm, after this sojourn in the barren solitude of the volcano. What pleasure to find again the shade of the pines, the flowers, and birds!

In 1846, the contrast became still more complete. We had passed the day on the side of Sorrento, in the beautiful country in which Tasso was born, and by nightfall we arrived on Vesuvius, then in eruption.

The explosions were not so violent as to prevent the crater being visited. A dark cloud crowned it, and was tinted with the reflection of the interior fire. We heard the noise of a formidable current issuing from the abyss, and streams of hot stones, launched out of sight, descended on the sides of the cone with a continuous crash. The eruption was most frequently preceded by a roll of thunder in the depths of the mountain; the earth trembled, and whilst the jet of gas penetrated the great mass of smoke, detonations like discharges of artillery shook the air. The lava could be seen to oose out through the cracks of the crater, and to run crackling in the inclined channels. On leaving the opening, the temperature decreased rather rapidly; the current is soon covered with scoriæ, which coalesce, and soon form a solid crust, on which one can pass without danger. At the centre the matter is still incandescent, after five or six years.

THE PHLEGRÆAN FIELDS.

Independently of the intermittent eruptive phenomena of Vesuvius, some permanent manifestations of a secondary order are noticed at its base, such as the numerous mineral sources of Castellamare, Santa Lucia, and the gaseous emanations of the sea near the Torre del Greco. A region situated to the west of Naples includes other important points connected with this great vent of subterranean fires. The ancients gave it the name of the Phlegræan Fields, and also that of the Forum of Vulcan.

Mythology makes this the scene of one of the labours of Hercules—his victory over the giants, 'sons of the earth,' thus symbolising the conquest of the fertile soil of that country during the period of repose which succeeded the eruptions of the early age. Over a surface of 100 square miles rises a series of hills in soft tufa, having the regular circular form which characterises the craters. Naples is built in the centre of a similar basin. Others are noticed at no great distance from the top of the Convent of the Camaldules, and from the promontory of Pausilippe. The little isle of Nisida is also an eruptive cone, with a crater open towards the sea-coast. The same circular arrangement is observed in the whole of the hills of Cumæ. Let us add the Isle of Procida, the group of the Ponza, the isles Ventotiene and Santo Stefano, which are evidently the remains of a larger island. Vesuvius besides is connected with the chain of extinguished volcanoes of Latium and Northern

Italy by two cones of large dimensions, Mont Vultur and Rocca Monfina, on the side of the Apennines, at mid-distance from the two seas. Vast craters are seen there, which have not been in activity during historical time, but which still give out some carbonic emanations.

THE SOLFATARA.

In the centre of the Phlegræan Fields, a very remarkable crater has received the name of the Solfatara, on account of the numerous sulphurous matters which it contains. It is situated near the town of Pozzuolo, the soil of which furnishes the volcanic product called pozzolane, so useful for all hydraulic constructions.

This basin, in which an eruption took place in 1198, still shows incontestable traces of the high temperature of subterraneous lava, with which a fissure doubtless puts it in communication. Hot sulphurous vapours constantly rise from the various orifices pierced through the bottom of the funnel, or in the volcanic rocks that surround it. In olden times they gave the name of *Leucogee*, or white hill, to the eminence formed by these rocks decomposed and whitened by the vapours. The minerals which are worked usually contain one-third of sulphur; nevertheless, in some parts this substance is naturally refined, and nearly pure. The rocks, dissolved by the rain, form, on the surface of the crater, beds similar to pipeclay, which the gases fill with bubbles.

It is thus that Mr. Poulett Scrope * accounts for the *rimbomba* which is heard at each step, and which other travellers have attributed to the abyss situated below the crust.

AVERNUS.

We cull the following from Lucretius :—'Now mark, and I will make clear to you what kind of nature the several Avernian places and lakes possess. First of all, as to the name Avernian, by which they are called, it has been given to them from their real nature, because they are noxious to all birds; for when they have arrived in flight just opposite to these spots, they forget to row with their wings, and fall, with soft neck outstretched, headlong to the earth, if so be that the nature of the ground admits of that; or into the water, if so be that a lake of Avernus spreads below. There is such a spot at Cumæ, where the mountains are charged with acid sulphur, and smoke, enriched with hot springs. Such a spot there also is within the Athenæan Walls, on the very summit of the citadel, beside the temple of bountiful Tritonian Pallas, which croaking crows never come near on the wing—no, not when the high altars smoke with offerings; so constantly they fly not before the sharp wrath of Pallas for the sake of your vigil kept, as the poets of Greece have sung; but the nature of the place suffices, by its own proper

* *Volcanoes, their Characteristics and Phenomena.* By G. Poulett Scrope.

power. In Syria too as well, a spot, we are told, is found to exist of such a sort that as soon as ever even four-footed beasts have entered in, its mere natural power forces them to fall down headlong, just as if they were felled in a moment as sacrifices to the manes gods. Now all these things go on by a natural law, and it is quite plain whence springs the cavern from which they are produced; that the gates of Orcus be not haply believed to exist in such spots, and next we imagining that the meaner gods from beneath do haply draw souls down from them to the borders of Acheron.' *

To the west of Pozzuolo several craters are filled with water, and form lakes. It is there that the Avernian Lake is found, to which the imagination of the ancients attached a gloomy legend. After the early eruptions, thick forests had covered the sides of the hills which surround it, and in the midst of their deep shade Homer places the abode of the Cimmerians. Virgil made it the terrible spot where the sybil conducts her heroes to the entrance of the infernal regions.

At the present time the forests have disappeared, and the bird fearlessly wings its way over a blue lake the borders of which are planted with vines. But deleterious gases must for a long time have escaped near this lake, as still occurs in many localities on the banks of the neighbouring lake of Agnano. At a short distance is the grotto called *Grotto del Cane*, in which small animals perish, while human beings

* Munro's *Lucretius*, vol. i. pp. 309, 310.

experience no inconvenience. This arises from the fact that in the grotto is a crevice, through which passes carbonic acid gas, which, being heavier than the air, forms a stratum near the floor, the deleterious influence of which only extends to a height of about two feet.

The vicinity of the Avernian Lake was much changed at the beginning of the sixteenth century by a violent eruption, during which Monte Nuovo was formed, of which we shall give a description further on. The *débris* thrown out by the crater partly filled Lake Lucrin. Violent earthquakes preceded the formation of this mountain, and calm was re-established after its appearance. Vesuvius was then in a state of inactivity, which lasted a century and a half.

Fig. 5.—ERUPTION OF ETNA IN 1754.

II

ETNA.

ANCIENT ERUPTIONS—THE VAL DEL BOVE—THE CRATER OF ETNA —ERUPTION OF 1669—ERUPTION OF 1865—EMPEDOCLES—THE CYCLOPS—EOLIAN ISLES—STROMBOLI.

ANCIENT ERUPTIONS.—THE VAL DEL BOVE.

WE read in the *Æneid* :*—

The flagging winds forsook us with the sun,
And, weary'd, on Cyclopean shores we run.
The port, capacious and secure from wind,
Is to the fort of thundering Ætna joined.
By turns a pitchy cloud she rolls on high,
By turns hot embers from her entrails fly,
And flakes of mountain flames that arch the sky.
Oft from her bowels massy rocks are thrown,
And, shivered by the force, come piece-meal down;
Oft liquid cakes of burning sulphur flow,
Fed from the fiery springs that burn below.
Enceladus, they say, transfixed by Jove,
With blasted limbs came trembling from above,
And when he fell, th' avenging father drew
This flaming hell, and on his body threw;
As often as he turns his weary sides
He shakes the solid hill, and smoke the heaven hides.

This description of Virgil's is one of the accounts which prove the activity of Etna during the centuries which preceded the Christian era. From that

* Book iii. ver. 570 and seq. Dryden's translation.

period the volcano passed through a long state of repose, but for the last 800 years violent eruptions have succeeded each other at short intervals, and their frequent return has multiplied the dislocations of the earth to such an extent, that at the present time more than 200 secondary beds can be counted on the sides of the mountain. The principal cone rises to a height of 3,608 yards above the sea, its top smoking and covered with snow. Among the ravines which intersect its sides, a deep valley—the celebrated Val del Bove—opens in the eastern side, and reaches as far as the sea. 'To me, I own,' says Mr. Poulett Scrope,* 'it has always appeared to have originated in a great fissure, enlarged into a crater by some paroxysmal eruption which blew out of the heart of the mountain, and since widened by the abrasive violence of aqueous *débâcles*, caused by the sudden melting of snows on the heights above by the heat communicated from erupted lavas and showers of red-hot scoriæ falling over the surface. One such torrent, indeed, is recorded, which rushed down the same valley in 1755, in the month of March, the volcano being at the time covered with snow; on which occasion the flood is said by Recupero to have run at the rate of a mile and a half in a minute for the distance of twelve miles—a rate which would give an enormous abrading and carrying force to a great body of water. Accordingly its track, two miles in breadth, is now visibly strewn to the depth of from thirty to forty feet with sand and fragments of

* *Volcanoes, their Phenomena, &c.*, p. 339.

rock. And that similar *débâcles* had previously for many ages taken the same course is demonstrated by the accumulation of a vast alluvial formation at the opening of the valley to the sea, near Giarre, more than 150 feet deep, measuring ten miles by three in area, and now resembling an upraised line of beach 400 feet high above the sea.'

During the great eruption of August 21, 1852, described by Sir Charles Lyell, a great number of openings from the summit to the base of the great precipice which forms the entrance of the valley showed themselves. From the cone made by the lowest opening flowed a large sheet of lava, which, precipitating itself in a cascade into a deep ravine, gave out in its fall a noise as of 'metallic substances and breaking glass.' This eruption lasted nine months, and the depth of the accumulated lava in some places was as much as fifty-four yards. The recent investigations of geologists have shown the immense quantity of materials added during eight centuries to the mass of Etna, and it may be imagined that, with all these accumulations, the entire mountain might be created by series of eruptions, succeeding each other for an indefinite period. This explanation, however, does not exclude the enlargement of the bulk by internal swelling, owing to the rising up of the strata, or to the injection of the lava in the crevices of the volcano.

CRATER OF ETNA.

M. Élie de Beaumont, who, accompanied by Leopold de Buch, ascended to the top of Etna in 1834, thus describes a crater in activity, which was situated on its summit:—

'It was to all of us a moment of surprise, rather difficult to depict, when we unexpectedly found ourselves not at the edge of the large crater, but of a circular gulf, from 90 to 100 yards in diameter, which only touches it at one point in its circumference. We looked eagerly down this nearly cylindrical funnel, but it was vain to seek there the secret of the action that was going on. The nearly horizontal strata, which were plainly seen on the almost vertical escarpments, only revealed to us the structure of the upper cone. Trying to count them, one under the other, they were gradually lost to view in the complete darkness of the bottom. No noise came from this gloomy depth; it exhaled only whitened vapours, slightly sulphurous, formed chiefly of steam. The dismal aspect of this black and silent abyss, in which our sight was lost; its obscure and damp sides, along which wound about, in a languid and monotonous way, long streams of vapour of a grey and melancholy tint; the large crater, to which is attached the narrow gulf, and in it the confused heap of differently coloured materials—some yellow, some grey, others red—appeared the picture of chaos. Everything around us presented a funereal and sepulchral aspect. The chill of the morning, seconded by a light north-east wind, heightened still more this sad and wild impression.'

The actual cone of Etna rises above a platform, the circular border of which marks the limit of an old crater much larger, filled since the time of its formation with lava and scoriæ. The configuration

Fig. 6.—Etna (1858) : the Crater.

of craters, as has been well observed by Mr. Poulett Scrope, is not only modified by the action of time and meteoric agents; the volcanic phenomena, by a series of alternating causes, producing in turn craters and deep excavations in the interior of the mountain, and afterwards filling them by eruptions, which at the same time increase the size of the newly-formed cones.

ERUPTION OF 1669.

This eruption, one of the most violent that is remembered, was preceded by a terrible hurricane, which began suddenly on March 8, two days before the beginning of the disaster. Some English merchants, who witnessed its different phases, have given a startling account of it:—

'Touching the forerunners of this fire, there was, for the space of eighteen days before it broke out, a very dark sky in those parts, with thunder and lightning and frequent concussions of the earth. . . . Besides, it was observed that the old top of Etna did, for two or three months before, rage more than usual, the like of which did Volcan and Strombilo, two burning islands to the westward. And the top of Etna must, about the same time, have sunk down into its old *vorago* or hole; in that 'tis agreed, by all that had seen this mountain before, that it was very much lowered. Other forerunners of this fire I have not heard nor met with.

'It first broke out on March 11, 1669, about two hours before night, and that on the south-east side, or skirt of the mountain, about twenty miles beneath the old mouth, and ten miles from Catania. . . . It overwhelmed in the upland country some fourteen towns and villages, whereof some were of good note, containing three or four thousand inhabitants, and stood in a very fruitful and pleasant country, where the fire had never made any devastation before; but now there is not so much as any

Fig. 7.—ERUPTION OF ETNA IN 1669. LAVA FLOWING OVER THE WALLS OF CATANIA.

sign where such towns have stood; only the church and steeple of one of them, which stood alone upon an high ground, does still appear.

'As to the matter which thus ran, it was nothing else but divers kinds of metals and minerals, rendered liquid by the fierceness of the fire in the bowels of the earth, boyling up and gushing forth like the water doth at the head of some great river; and having run in a full body for a good stone's cast or more, the extremities thereof begun to crust and curdle, becoming, when cold, those hard porous stones which the people call *Sciarri*, having the nearest resemblance to huge cakes of sea-coal full of a fierce fire. They came rolling and tumbling over one another, by their weight bearing down any common building, and burning up what was combustible. The chief motion of this matter was forward, but it was also dilating itself, as a floud of water would doe on even ground, thrusting out severall armes, or tongues, as they call them.

'About two or three of the clock in the night we mounted an high tower in Catania, whence we had a full view of the mouth, which was a terrible sight, viz. to see so great a mass or body of meer fire. Next morning we would have gone up to the mouth itself, but durst not come nearer than a furlong off, for feare of being overwhelmed by a sudden turn of the wind, which carried up into the air some of that vast pillar of ashes, which, to our apprehension, exceeded twice the bigness of *Paul's* steeple *in London*, and went up in a streight body to a far greater hight than that; the whole air being thereabout all covered with the lightest of those ashes, blown off the top of the pillar. And from the first breaking forth of the fire untill its fury ceased (being fifty-four days), neither sun nor starr were seen in all that part. . . . .

'At the mouth whence issued the fire or ashes, or both, was a continual noise, like the beating of great waves of the sea against rocks, or like thunder afarr off, which sometimes I have heard here in Messina, though situated at the foot of high hills and sixty miles off. It hath also been heard 100 miles northwards of this place, in Calabria.

'About the middle of May we made another journey thither, where we found the face of things much altered; the city of Catania being, three-quarters of it, compassed round with these sciarri as high as the tops of the walls. . . . .

'The general face of these sciarri is, in some respect, not much unlike, from the beginning to the end, to the river of Thames in a great frost at the top of the ice above bridge—I mean, lying after such a rugged manner in great flakes; but its colour is quite different, being most of a dark dusky blew, and some stones or rocks of a vast bigness, close and solid.'*

During this eruption the lava accumulated before the walls of Catania, sixty feet high, poured over them without levelling them down, and there is still to be seen an 'arcade of lava curling over the wall

* *Philosophical Transactions*, vol. iv. 1669, p. 1028–1031.

Fig. 8.—ETNA. CASCADE OF FIRE ; ERUPTION OF 1771.

like a wave on the beach.'* This curious phenomenon, and other analogous facts, show that the vapour which escapes from the surface of the lava current accumulates between this surface and the level surface of the obstacle, and prevents immediate contact. The lava is then seen to stop, as if by magic, some inches from the resisting surface, which should be of such a size that the vapour may fill the narrow intermediate space, and oppose a sufficient force to the slow movement of the lava.

If the impulsion of the lava current is considerable the obstacles give way; and it was thus that, during the same eruption of 1669, the large torrent which descended on Catania, caused grave disasters, and formed a promontory which stretches for more than half a mile into the sea. This torrent rose at the lowest point of an enormous fissure opened in the south-east flank of Etna. It covered an area thirteen miles long, and six wide in some places. Happily met in its passage by another current, which went westward, it was turned from its path, and, bordering the ramparts of Catania, it passed the port, and reached the sea on April 23. Then commenced, between the fire and the water, a combat, of which it is impossible to form an idea, but which those who witnessed the terrible scene gave up describing. The lava, cooled at its base by the contact of the water, presented a perpendicular front, fourten or fifteeen yards broad, and advanced slowly, carrying, like a glacier, enormous solid blocks still

* Poulett Scrope.

red with heat. When they reached the extremity of this movable causeway the blocks fell into the sea, filled it little by little, and the fluid mass advanced by so much. At this contact large bodies of water, reduced to a state of vapour, rose up with frightful hissings, and fell in a salt rain over the neighbouring country. In some days the lava had carried forward the line of the beach some 330 yards.*

During the eruptions of 1754, 1766, 1771, 1780, 1792, 1809, and 1812, enormous rents have opened passages for the lava and scoriæ, by orifices situated in the line of fissure. Some of the cones, formed in a few days round these orifices, measured 1,000 feet in height. The lava beds are, on the average, 26 to 32 feet thick, and sometimes much more. We see then, clearly, that the mass of Etna must have been wonderfully increased by this enormous quantity of material ejected during historic time; and in comparing this period with the primitive one, the immense length of which is shown by geology, we are led to believe that the greater part of the mountain, like Iceland, has been formed by a series of eruptions and by the interior accession of injected lava.

ERUPTION OF 1865.

Frequent shocks of earthquakes, felt on the sides of Etna in the month of October 1864, were the first symptoms of this recent eruption. Towards the end of January a whirlwind of smoke rose from the

* *Souvenirs d'un naturaliste.* By A. de Quatrefages.

Fig. 9.—ETNA. ERUPTION OF 1766.

crater, and, at the same time, dull roarings, accompanied by slight shocks, were heard on the eastern slope of the Apennines. Round Etna the atmosphere, although calm, was suffocating; the column of smoke continued to rise higher and to become more dense; and other well-known signs portended a coming awakening of the volcano.

During the night, from January 30 to 31, a violent shock made the inhabitants leave their houses, in the villages situated on the north-east side of the mountain.* Immediately after sheaves of fire rose in one place, about 1,800 yards above the sea, and as soon as the ground opened the lava began to flow rapidly: in two or three days it had extended in length three miles, with a breadth of one or two; its depth was variable, but often attained from thirty-two to sixty-five feet. This stream of lava, checked by an ancient cone, divided into two arms, one of which precipitated itself into a straight and deep valley, forming a cascade of fire, carrying on its surface solidified blocks, which fell with great noise from a height of 50 yards. The craters, to the date of March 10, were seven in number, five of which were comprised in a vast crevassed belt, closed on all sides except towards the west, where there was an opening from which the torrents of lava escaped. These craters were situated on the prolongation of a large rent in the surface, probably produced at the begin-

* We owe these details to an interesting letter addressed by a geologist, M. Fouqué, a witness of the eruption, to M. Ch. Sainte-Claire Deville.

ning of the eruption—a fact frequently observed, as may be seen from the preceding accounts.

'The three upper craters,' says M. Fouqué, 'give forth, about two or three times a minute, very strong detonations, resembling the rolling of thunder. The lower craters, on the contrary, give out noises so

Fig. 10.—Eruption of Etna in 1865: the Crater before Eruption.

unceasingly that it is impossible to count them. These noises succeed one another without rest or repose; they are sharp, and distinct one from the other. I cannot do better than compare them to the strokes of a hammer falling on an anvil. If the ancients heard similar noises, I can very well conceive the idea which it gave them—that of a forge in the centre of Etna, with the Cyclops as workmen.'

EMPEDOCLES.

The fertility of volcanic lands does not show itself with greater advantage in any part than in the plains and beautiful valleys of Sicily, at the foot of the fertile slopes of Etna, planted with vines, olives, pines, chestnuts, and oaks. 'The earth,' says Homer, 'is fertile without being sown or cultivated; it produces wheat, barley, the vine—the grapes of which give abundant wine—and the rain of Jupiter makes the fruits grow.' It was in the midst of the rich harvests of this magnificent region that Ceres had her first altars. Sicily was formerly called the granary of Rome, which not only drew from it abundant fruits, but also found in its great men, the soul of Greece, inexhaustible sources of civilisation and progress. We cannot here trace the history of the philosophers and illustrious men of this flourishing island, during the government of the Greeks; we shall only speak of Empedocles, born at Agrigente, who excelled in the sciences, philosophy, poetry, and music. Descended from one of the first families of Sicily—handsome, eloquent, generous—he refused the royalty which was offered to him; and, guided by higher ambition, tried to reform manners, and to contribute to the good of his country in helping forward the development of its liberties. A disciple of Pythagoras, he believed that unity is the principle of all things, and contested the existence of the gods of his time. Accused of pride and impiety, he had to submit to the attacks of calumny; and, after his death, the report was that he had precipitated himself into the crater of Etna, in order to spread belief in his apotheosis and to obtain divine honours. But a brass sandal thrown out by the volcano, which was recognised to belong to him, unmasked his strange vanity. If we believe this account, it is infinitely more probable that Empedocles perished, like Pliny, a victim to his zeal for science, and that the unlikely fact of his voluntary death was suggested by the enemies, who had been made such by his high sense and courageous sincerity.

The fame of Empedocles was very great. The verses of his poem on Nature, of which only a few fragments are left us, were chanted in the Olympian games, and his numerous works on science and morals had obtained for him the admiration of the best minds of his time. The inhabitants of Agrigente, anticipating the homage of posterity, erected a statue to his memory. The Sicilian legends place his residence in a ruined tower, the *Torre del Filosofo*, an ancient monument erected on the precipitate peaks which dominate the *Val del Bove*. In the fine painting of Rafaelle, representing the School of Athens, one of the most magnificent works which art has left to us, Empedocles is placed between Archimedes and Pythagoras.

THE CYCLOPES.

The Cyclops, according to the legend, had established their forges in the caverns surrounding Etna, the crater of which, says Pindar, resembles a large furnace chimney. Ancient powers of the air, Vulcan's blacksmiths, god of fire, they manufactured celestial thunderbolts, and the noise of their hammers was heard from afar. Later on they made arms for mortals, and they were thus confounded with the mysterious Kabires, priests of Ceres, cunning metallurgists, who may be considered as the first industrial powers. It was on the volcanic heights of the Caucasus that Prometheus, inventor of fire, creator of the arts and of industry, was chained.

It is also certain that the hard stones proceeding from the igneous rocks, and rolled by the sea or by torrents, served in old time for hammers or arms, according to their size and form, similar to those now made by the different peoples of the volcanic islands of Oceania.

Enceladus, the most powerful of the Titans buried under Sicily, according to the fable, was the cause of the eruptions of Etna. 'Every time the giant moved he caused flames to ascend, or gave rise to disturbances on the earth or in the waters.' M. Élisée Reclus adds, on this subject:—

'One cannot help contemplating the volcano as if it were a being endowed with an individual life, and rejoiced in the consciousness of its strength. The features of Etna, so regular and noble in their repose, resemble somewhat the figure of a sleeping god; it is not so much, as the old legend pictures it, the mountain which weighs on the body of Enceladus—it is the Titan himself, the ancient divine protector of the Sicilies, forsaken for the younger gods of Greece, the masters of Olympus.' *

Theocrites, in one of his idyls, makes Polyphemus thus describe the pleasant country of Sicily: 'There are laurels, tall cypresses, dark ivy, a vine with sweet fruits, and fresh water—ambrosial liquid, which Etna sends me from its white snows.'

The great fertility of volcanic lands, of which we have already given an example in the Phlegræan Fields, would doubtless cause to reside there, at the origin of communities, the primitive races of whom the legendary history is confounded with fable, which indicates to us, nevertheless, by striking features, the first efforts—the first conquests—of man seeking for and discovering in Nature the means to spread his dominion, and to establish on a more solid basis a well-being, at first too unstable, too much subjected to the formidable condition of the soil and the atmosphere. Doubtless the vicinity of the craters, although they afforded him a more fertile soil, placed him under the direct influence of disastrous commotions; but all those who have inhabited the beautiful countries which surround the still active volcanoes, will understand the strange charm of these poetical regions, where we seem to see the life, so to speak, of the planet which holds us, and where the first men believed themselves surrounded by the most marvellous manifestations of the power of the gods.

* *Revue des Deux Mondes*, July 1, 1865.

THE EOLIAN ISLES.

These small islands, seven in number, situated near Sicily, and now called the Lipari Isles, are the ancient Vulcaniæ on which reigned Eolus, god of Winds. The mountain of Hiera was consecrated to Vulcan, who had his palace in the island. We see by these fables, that the Eolian group was celebrated both for its volcanoes, and on account of the very frequent meteoric phenomena in this part of the Mediterranean. Lying between Naples and Sicily, these islands may be considered as belonging to a single system, to a single submarine volcano, at the present time active at the orifices of Stromboli (Strongyle) and of Vulcano (Hiera). Before describing these two volcanoes, we will take a glance at the whole of the islands, which Mr. Poulett Scrope 'strongly recommends to the study of geologists who desire to form an opinion for themselves on volcanic phenomena.'

The ancient 'Forges of Vulcan' all have craters on their summits. The Phenician navigators, struck by the noise that came from them, which they heard at a great distance, gave them the name of Musicians' Isles. The mountains which form them are nearly entirely composed of enormous accumulations of lava, the alternate strata of which attain a thickness of several hundred feet. At Ustica the presence of marine shells in the strata prove the recent emergence of the island. Hot water springs spout up from the sides of the hills, and abundant gaseous emanations decompose the materials of which the craters are composed.

M. Ch. Sainte-Claire Deville, in a letter addressed from Naples to M. Dumas, on November 3, 1855, thus describes the Eolian Isles :—

'The three principal isles, Lipari, Vulcano, Stromboli, each possesses a particular interest. The two last especially present to the geologist the most precious teachings. I advanced into the crater of Stromboli as far as I could do without rashness. Like my predecessors in this place, Spallanzani and Poulett Scrope, I was abandoned by my guides, and was obliged to advance alone; but I was fully recompensed. From the point which I reached, the eyesight plunged nearly vertically into the chimney where Spallanzani saw the lava, in a state of fusion, alternately rise and descend; and I distinguished, at a few yards' distance, the bed whence shot forth, about every ten minutes, and with a noise which, at this slight distance, was of a most striking character, a column of vapour, drawing with it, to a great height, incandescent stones, which fell again, partly in the mouth itself and partly on the outer slope. Nevertheless, the vapour, which was driven by a north-west wind, considerably spoiled my observations, and I cannot too strongly recommend to geologists to choose a wind from the south, in order to enjoy the spectacle well.

'Vulcano is perhaps the most curious volcanic point in the Mediterranean. It presents a double interest: it is one of the most perfect volcanoes of upheaval that can be seen; from the point of view of chemical geology, it is the most beautiful solfatara which exists. . . . There is no more striking spectacle than that presented at night by the bottom of this immense funnel, whence the bluish flame of sulphur in combustion is seen to arise from a great number of small holes situated at the foot and on the surface of a hillock.'

STROMBOLI.

Stromboli, noticed by Homer, and which still serves as a lighthouse to mariners, has been in full activity since the most remote periods. It throws out flames continually without being actually in eruption, although the nature of the ground shows that formerly it was subject to them. The crater situated at the top of the island is notched towards the north, and on the same side scoriæ flow into the sea by a very inclined slope. The observations made by geologists who, since Spallanzani, have visited this curious perpetually-active volcano, throw great light on the phenomena of eruptions, which are difficult to study under ordinary circumstances. The crater of Stromboli is very favourable to this enquiry. Easy of access in all seasons, 'one is in a sort of way admitted,' as Mr. Scrope remarks, 'into the secrets of the laboratory of Nature, which is here open to our minute examinations.'

We have already quoted the description of M. Ch.

Sainte-Claire Deville. Those of Spallanzani, of Poulett Scrope, and of the German geologist Friedrich Hoffmann, one of the latest explorers, differ very little. They present to us the lava under the form of a shining mass, as of molten metal, brilliant even with a glittering light in full day, which, from one quarter of an hour to another, rises with a dull

Fig. 11.—Stromboli.

grumbling to the edge of the crater, opens at its centre with noise, making the earth tremble, and vomiting, in the explosion, a sheet of incandescent lava and burning scoriæ. 'The surface of the lava,' says M. F. Hoffmann, 'rose and fell regularly at rhythmical intervals. A particular noise is heard, similar to the decrepitations of the air entering by the door of a mine-furnace. A spherical mass of white vapour leaves it at each decrepitation, lifting up the lava, which again falls, after its leaving. These shells of vapour draw from the surface of the lava red-hot scoriæ, which dance as if movèd by invisible hands above the edge of the opening.'

Similar appearances have been observed in the volcanoes of Masaya and Bourbon, in the crater of Kilauea at Hawaii, as well as in those of Vesuvius and Etna. When the eruption is permanent, there certainly exists below the orifices a mass of liquid lava in constant ebullition, which furnishes the prodigious mass of matters ejected during centuries, and which must be replenished by a cause still unknown, but probably analogous to that which connects the volcanoes of the same chain, often at great distances. It is, moreover, by admitting such affinities, that we can comprehend the action of volcanic mountains, which, as has been very well said by Seneca, even do not furnish the aliment of fire, but only give it an issue.

In the permanent volcanoes, the force of subterranean expansion produces, as may be imagined, effects more or less energetic according as the weight of the atmosphere, the principal force of repression, increases or diminishes. Indeed the inhabitants of Stromboli, who are mostly fishermen, habitually observe the phenomena of this volcano, in order better to forecast atmospheric variations. During the storms of winter, the eruptions are sometimes very violent, as we have observed in passing near the volcano at those times. These eruptions split the sides of the crater in the very act of explosions which shake all the island, and make themselves heard to a great distance. At Ternate in the Moluccas, and in many other volcanic regions, the same coincidence has been observed between eruptions and tempests.

III

ICELAND.

ERUPTIONS OF HECLA AND KOTLUGAIA—OVERFLOWING OF LAVA—ERUPTIONS OF SKAPTAR-JÖKULL—THE GEYSERS—NATURAL CURIOSITIES—FORMATION OF ICELAND BY VOLCANIC ERUPTIONS—STREAMS OF LAVA.

ERUPTIONS OF HECLA AND KOTLUGAIA.

THE desolate aspect of Iceland is owing as much to the enormous masses of volcanic *débris*, which cover the surface, as to the immense glaciers that come down the sterile mountains, and to the marshy plains where perhaps ancient forests rose, the remains of which are pointed out by the inhabitants. A considerable sinking of the temperature in the northern regions, which tradition proves to have been in olden times more peopled and flourishing, is not the only cause of the rude conditions of existence in this isle at the present time. Since the year 1000, nearly fifty great eruptions have taken place, some of which have inundated the surface of the country with lava, cinders, and scoriæ, and decimated the population.

Mount Hecla, situated in the southern part of the island, at a little distance from the shore, is especially remarkable for the frequency of its eruptions, which have often coincided with those of Vesuvius and Etna.

The violent paroxysm of 1766 covered the surrounding country in a thick bed of *débris*. The rain of cinders reached to a distance of 149 miles, and the air was so darkened that objects could not be distinguished in many parts of the island. Shortly after, a torrent of lava flowed from the crater, and was soon followed by an immense column of water, which added its ravages to those of the fiery eruption.

In 1845, the top of the volcano was shattered by an explosion, and the mountain lost 500 feet of its height. The stream of lava in this last eruption reached a distance of nine miles, its thickness varying from 40 to 80 feet. Although enormous streams of lava have thus covered a great portion of the soil of Iceland, the showers of cinders and scoriæ proceeding from the numerous craters which have been in eruption since the historic period, appear to have especially sterilised the country, to which no other region of Europe can be compared for volcanic activity. In these high latitudes, each eruption causes also the melting of enormous masses of snow and ice, and it follows that 'a large proportion of the formations of Iceland consists of conglomerates formed by the tumultuous rush of floods from the eruptive heights, carrying along enormous quantities of alluvial matter, which they spread in wild confusion over the lowest levels, filling up some valleys and excavating others.' . . . .

'During the eruption of Kotlugaia in 1756, prodi-

Fig. 12.—MOUNT HECLA.

gious torrents of water, ice, rocks, and sand, occasioned by the melting of the glaciers, rushed from the heights, and produced three parallel promontories, reaching several leagues into the sea, which remain above its level in places where the fishermen formerly found forty fathoms of water.' *

The crater of Kotlugaia is an immense fissure which crosses the mountain, split in two during an eruption. The snows, ice, and smoke bar an approach to this abyss, which is distinctly visible at a distance of sixty-five miles. Lyell quotes the fact of a bed of ice found under the lava of Etna, and preserved by a bed of sand and scoriæ. The same thing often occurs in Iceland, and it follows that the transmission of heat to these glaciers, which uphold beds of rock and volcanic matter, precipitates down the sides torrents of water and *débris*. These floods draw with them all that they encounter on their passage. Entire forests are thus buried at the base of the volcanoes, and form beds of fossil wood, which are often found there, and which in Iceland, as we have already said, seem to indicate a more favourable climate in former periods. In the recent eruption of Kotlugaia, in May 1860, torrents of water again accompanied a column of black vapour and fiery scoriæ, which rose to a height of 7,600 yards, and the melting of the ice again caused the carrying away of enormous masses of rock, which were transported as far as the sea.

'The devastating effect of such sudden floods may

* *Volcanoes, their Characteristics and Phenomena, &c.*, by G. Poulett Scrope, p. 172.

well be conceived: they not only heap up vast volumes of conglomerate on the plains below, but also tear up and score the mountain with ravines of proportionate size—groove, striate, and polish its hardest rocks with the rolling ice and stony flood, and add miles of new land to the coast line. When we add to this the dense showers of scoriæ and ash that fall continuously for days together from the air, into which they are ejected by the volcanoes, and the torrents of seething lava that, issuing from its entrails, rush down the slopes, in conjunction with the ice and water *débâcles*, and cover many square miles of surface with sheets of solid rock, it is evident that few more powerful agents of superficial change can be imagined among all the living forces of nature.' *

These agents, which we now see at work, have powerfully contributed to the formation of the crust of the globe, at epochs when the energetic commotion of the elements increased the number and grandeur of the phenomena. In the midst of commotions, subversions, and changes, brought about by the two principles of water and fire, the repose of the immense mass, of which all the materials proceed from the prodigious struggle of the primitive forces, was prepared in an apparent disorder. These forces, still at work, but with less energy and to a less extent, are but the consequence of the laws which have presided, and which still preside, over the organisation of the material world, the cradle of humanity. The observations of geologists go to

* Scrope, *loc. cit.* p. 409.

Fig. 13.—CRATER OF HECLA.

prove that Iceland has been entirely uplifted from the bottom of the sea 'by successive eruptions of a like volcanic system.' Mr. Poulett Scrope believes even that it can be said, 'from the same volcano,' and the arrangement of the orifices in activity support his opinion.

OVERFLOWING OF LAVA.

Among the phenomena which the interior of the island presents to the curiosity of naturalists, one of the most remarkable is the valley of Thingvalla, formed by the sinking of the upper and central part of an immense overflowing of lava, emitted from the foot of the mountain of Hrafnabjorg, or, according to the record, from the middle of the ancient forest of that name. This enormous sinking has left on each side of the valley, the length of which is about four miles, and the depth 800 feet, a sudden precipice, a sort of abyss, the height nearly reaching 200 feet. The most imposing of these gigantic fissures, called Almanna Gja, is about two or three leagues in extent, and at a distance is like an immense fortification. In some points, the lower beds of lava are arched, as if to resist the pressure of the upper beds, and have formed vaults similar to those of our large buildings. The river Oxeraa, which descends in torrents in the crevasses of the plateau of Almanna Gja, called also the Hill of Laws, reunites its waters in a dark and deep basin bordered with crumbling rocks, where formerly women were thrown over when condemned to death for adultery.

The volcanic chasm on the opposite side, called Rabna Gja, equally extensive, but not so deep, and fallen down in places, does not give so clear an idea of the prodigious catastrophe of which Thingvalla has been the theatre.

Innumerable fissures cut the bottom of the valley,

Fig. 14.—Crevasses of Almanna Gja.

and two amongst them, which reunite, enclose an oval piece of land, which can only be entered by a narrow passage, where the old Scandinavian parliament of the island, the *Althing*, each year in July, held its great meetings, at a time when Europe was still subjected to the despotism of feudal government. The president of these free assemblies, named Logmadr, man of the law, was chosen for life by the people. It was in this place, one of the most celebrated in Iceland, that Christianity was adopted, by a majority of voices, in the year 1000.

A magnificent lake, the water as transparent as the bright tints of the emerald, rests at the bottom of the plain of Thingvalla, clothed in the fine season

Fig. 15.—The Lake of Thingvalla.

with grass and shrubs. We quote from a very interesting account of a voyage in the North Seas* the following description of this strange and magnificent place:—

'A blaze of light smote the face of one cliff, while the other lay in the deepest shadow; and on the rugged surface of each might still be traced corresponding articulations, that once had dovetailed into

* *Letters from High Latitudes,* by Lord Dufferin, fifth edition, p. 77.

each other, ere the igneous mass was rent asunder. So unchanged, so recent seemed the vestiges of the convulsion, that I felt as if I had been admitted to witness one of Nature's grandest and most violent operations almost in the very act of its execution. A walk of about twenty minutes brought us to the borders of the lake—a glorious expanse of water, fifteen miles long, by eight miles broad, occupying a basin formed by the same hills, which must also, I imagine, have arrested the further progress of the lava torrent. A lovelier scene I have seldom witnessed. In the foreground lay huge masses of rock and lava, tossed about like the ruins of a world, and washed by waters as bright and green as polished malachite. Beyond, a bevy of distant mountains, robed by the transparent atmosphere in tints unknown to Europe, peeped over each other's shoulders into the silver mirror at their feet; while here and there, from among their purple ridges, columns of white vapour rose like altar smoke toward the tranquil heaven.'

ERUPTION OF THE SKAPTA-JÖKUL.

The volcanoes, or jökuls,* of Iceland are situated on two parallel lines, crossing the island from north-east to south-west, and having between them a deep fissure, which gives birth to the immense quantities of lava with which Hecla, Kotlugaia, Sneifels, Skapta, &c., are surrounded. In 1783, this last volcano sent

* The Icelanders give this name to all high mountains constantly covered with snow.

forth two enormous torrents, which spread to a distance of from forty to fifty miles, with a breadth of seven to fourteen. The depth of the lava was in places 150 yards, and it was calculated that the mass deposited by this single emission—one of the most considerable which has been known, exceeded the volume of Mont Blanc. Lava spouted out of various sources opened at the foot of Skapta-Jökul, and situated in the direction of a fissure formed by the pressure of the igneous matter below. On the prolongation of this line, at a distance of thirty miles, and during the eruption, an island, which has since disappeared, suddenly rose from the sea. The Skapta-Jökul rises from a vast inaccessible space—a desert of lava and ice, whence flows the most dreadful plague which has ravaged the island.

'This event occurred in the year 1783. The preceding winter and spring had been unusually mild. Toward the end of May, a light bluish fog began to float along the confines of the untrodden tracts of Skapta, accompanied in the beginning of June by a great trembling of the earth. On the 8th of that month, immense pillars of smoke collected over the hill country towards the north, and, coming down against the wind in a southerly direction, enveloped the whole district of Sida in darkness. A whirlwind of ashes then swept over the face of the country, and on the 10th innumerable fire spouts were seen leaping and flaring amid the icy hollows of the mountain; while the river Skapta, one of the largest in the island, having first rolled down to the plain a vast volume of fetid waters mixed with sand, suddenly disappeared.

'Two days afterwards a stream of lava, issuing from sources to which no one has ever been able to penetrate, came sliding down the bed of the dried-up river, and in a little time—though the channel was six hundred feet deep and two hundred broad—the glowing deluge overflowed its banks, crossed the low country of Medalland, ripping the turf up before it like a table-cloth, and poured into a great lake, whose affrighted waters flew hissing and screaming into the air at the approach of the fiery intruder. Within a few more days, the basin of the lake itself was completely filled, and having separated into two streams, the unexhausted torrent again recommenced its march; in one direction overflowing some ancient lava fields; in the other, re-entering the channel of the Skapta, and leaping down the lofty cataract of Stapafoss. But this was not all: while one lava flood had chosen the Skapta for its bed, another, descending in a different direction, was working like ruin within and on either side the banks of the Hverfisfliot, rushing into the plain, by all accounts, with even greater fury and velocity. Whether the two issued from the same crater it is impossible to say, as the sources of both were far away within the heart of the unapproachable desert, and even the extent of the lava flow can only be measured from the spot where it entered the inhabited districts. The stream which flowed down Skapta is calculated to be about fifty miles in length by twelve or fifteen at its greatest breadth; that which rolled down the Hverfisfliot, at forty miles in length by seven in breadth. Where it was imprisoned between the high banks of Skapta, the lava is five or six hundred feet thick; but as soon as it spread out into the plain, its depth never exceeded one hundred feet. The eruption of sand, ashes, pumice, and lava continued till the end of August, when the plutonic drama concluded with a violent earthquake.'*

During a whole year, says Arago, at the end of this eruption, the atmosphere of Iceland was mixed with clouds of dust through which the rays of the sun scarcely penetrated.

THE GEYSERS.

'I must now direct your attention to a natural steam-engine which long held a place among the wonders of the world. I allude to the Great Geyser of Iceland. The surface of Iceland gradually slopes from the coast towards the centre, where the general level is about 2,000 feet above the sea. On this, as a pedestal, are planted the Jökull or icy mountains, which extend both ways in a north-easterly direction. Along this chain occur the active volcanoes of the island, and the thermal springs follow the same direction. From the ridges and chasms which diverge from the mountains enormous masses of steam issue at intervals hissing and roaring; and when the escape occurs at the mouth

* *Letters from High Latitudes*, ed. cit., p. 82.

of a cavern, the resonance of the cave often raises the sound to the loudness of thunder. Lower down in the more porous strata we have smoking mud pools, where a repulsive blue-black aluminous paste is boiled, rising at times in huge bubbles, which, on bursting, scatter their slimy spray to a height of fifteen or twenty feet. From the bases of the hills upwards extend the glaciers, and above these are the snow-fields which crown the summits. From the arches and fissures of the glacier vast masses of water issue, falling at times in cascades over walls of ice, and spreading for miles over the country before they find definite outlet. Extensive morasses are thus formed, which add their comfortless monotony to the dismal scene already before the traveller's eye. Intercepted by cracks and fissures of the land, a portion of this water finds its way to the heated rocks underneath; and here, meeting with the volcanic gases which traverse these underground regions, both travel together, to issue at the first convenient opportunity, either as an eruption of steam or a boiling spring.' *

The most famous of these boiling springs, situated to the south of the island, is the Great Geyser (a word which also signifies fury in the Icelandic idiom). It is a tube of 82 feet in depth and 9 feet in diameter, surmounted by a basin which measures 52 feet from north to south, and 59 from east to west.

'The eruption begins by a trembling of the earth,

* Tyndall, *On Heat considered as a Mode of Motion*, p. 119.

Fig. 16.—THE GREAT GEYSER.

in the bosom of which are apparently heard dull discharges of artillery. The observer thus warned has nearly always time to approach to a little distance from the basin, and can even stand on the slight decline that forms the cone, where he observes great commotions every time the liquid column rises. At first the waters overflow with a remarkable noise, owing, doubtless, to the sharpness of the prominences which clothe the cone. Some time after enormous bubbles of water are seen at the surface, which after having attained two or three feet in height suddenly stop. All is again quiet. This is a false eruption, which sometimes takes place two or three times running. But when the phenomenon is seen in all its majesty, jets that rise higher succeed to the bubbles of which I have just spoken, until they reach a height of from eight to ten feet. Then, as in fireworks, when all seems finished, the bouquet all at once comes which fills the observer with admiration; so it is with the Geyser, which, after a few minutes' rest, appears to gather all its strength and by a last effort throws up into the air an immense sheaf of water, the highest point of which generally attains a hundred feet. An enormous mass of white vapour afterwards hovers for some time above this imposing scene. The Geyser, whose fury is abruptly stopped, is again filled slowly and begins to flow afresh like a simple spring.'* A silicious

* *Voyage en Islande et au Groenland,* sur la corvette La Recherche, commandée par M. Tréhouart, publié sous la direction de M. Paul Gaimard.

bed, very smooth and hard, comparable to the most beautiful stucco, lines the interior of the tube and basin which holds the spring, the incessant work of which has created this wonderful apparatus. The deposit of silica which is formed on the adjacent parts of the edge gets slowly higher, making the

Fig. 17.—Basin of the Great Geyser.

well of the geyser deeper, constructing the mound at the top from which it springs. Its waters, clear as crystal, are inodorous, and have no disagreeable flavour; when cold they can be drunk with pleasure.

The Strockr, situated at about fifty paces from the Great Geyser, constantly makes a very loud bubbling, which has procured for it the epithet of 'the Devil's Kettle.' Its eruptions can generally be induced by throwing in clods of earth or grass. Then, without any perceptible moving of the earth, and after having completely suspended the tremendous noise that it makes, it is seen slowly rising up from the kind of well that it occupies, and scarcely has it reached the edge when, like a furious animal which has been forced to leave its retreat, it darts into the air a magnificent column, and scalds the face of the individual who has been imprudent enough to provoke the display.

At first the waters, coloured by the clods of earth with which its reservoir has been filled, are for some time blackish like mud, which singularly contrasts with the limpid stream of the Great Geyser, but it does not take long to regain its ordinary colour. In one of our experiments, a gun charged with shot, and fired from the surface, instantly stopped the boiling of the Strockr. Twenty minutes afterwards a fresh quantity of earth was thrown in. It again filled its funnel, without any disturbance of the ground, and immediately began to spout forth, with extreme violence, to a height which equalled the highest ascents of the Great Geyser. We excited it in a remarkable manner by continuing to throw in earth, and especially by shots. I think that the activity of this geyser, and the height of its jets, is owing particularly to its slight breadth and the length of its tube. Towards the end of this phenomenon, which had filled us with admiration, the Great Geyser was heard. The Strockr immediately returned to its well, but in another instance the smaller one quieted the larger. During these curious and imposing experiments, which seemed to prove an intimate connection between the two largest geysers of Iceland, the other basins full of water, although situated above and near them, never changed their level. Lastly, I will add that from

Fig. 18.—Strockr and Boiling Springs.

noon until eight in the evening we could work the Strockr several times without being able to exhaust it, for the water was ejected, the last time we excited it, nearly on our tent, and to a much greater distance than the preceding ones. It appeared to have redoubled in fury.*

The detonations and explosions of the geysers are

* E. Robert, *Voyage de la Recherche.*

explained by the production of the vapour in the pipes which feed the tube. We owe to M. Bunsen a beautiful theory of these eruptions, reproduced in Professor Tyndall's excellent work.

The variations in the jets of the geysers have no fixed time. When they occur very strongly, the Icelanders look for rainy and windy weather.

Round these principal springs, at a distance at most of half a mile, fifty boiling fountains may be counted. These probably have all the same origin. The water is generally clear, but in crossing the veins of ochre and whitish clay, it sometimes becomes red, like blood, or takes the colour of milk.

The geysers of the north, like those of the south, occupy the bottom of a large fissured valley. The most important throws a jet of about five yards high. It is situated between two others, which furnish a much larger quantity of water, but without eruption, and which thus takes the name of *laug* (bath).

'Some of those [laugs] in Iceland are forty feet deep. Their beauty, according to Bunsen, is indescribable; over the surface curls a light vapour, the water is of the purest azure, and tints with its lovely hue the fantastic incrustations on the cistern walls; while at the bottom is often seen the mouth of the once mighty geyser. There are in Iceland traces of vast, but now extinct, geyser operations. Mounds are observed whose shafts are filled with rubbish, the water having forced a passage underneath, and retired to other scenes of action. We have, in fact, the geyser in its youth, manhood, old age, and death, here presented to us—in its youth, as a simple thermal spring; in its manhood, as an eruptive column; in its old age, as the tranquil laug; while its death is recorded by the ruined shaft and mound which testify the fact of its once active existence.'*

NATURAL CURIOSITIES.

Iceland, which may be justly named 'the queen of volcanic islands,' contains an infinite number of natural curiosities produced by the double action of its volcanoes and its immense glaciers which crown their summits.

In some regions the country presents the most strange and wonderful aspect. Fogs, so frequent in these parts, often have a reddish tinge, and M. E. Robert, in his picturesque account, says that they may often be taken for a shower of volcanic ashes. The violent action of the winds also produces remarkable effects. During the drought, magnificent waterspouts and immense clouds of red dust are suspended very high in the atmosphere. They sometimes remain there for a long time after calm has been restored below, and are carried to great distances out to sea. These clouds in passing discolour the lower part of the snow that clothes the mountains, and, at these times, it is possible to fancy oneself in the midst of a volcanic eruption. This phenomenon, so remarkable and rare, is called *Mistur* in Iceland.

If the atmosphere is so singularly connected with

* Tyndall, *On Heat*, p. 126.

the volcanic phenomena, the soil of the island retains the most striking traces of the action of subterranean fires. The mountains of Esia, situated near Reykiavick, the capital of Iceland, at a distance appear covered with flourishing vegetation. Their steep arid sides owe this appearance to the beautiful green tint of the rocks which compose the greater part of the chain; their upper strata present a great variety of colours. Mount Husaell, in the valley of Reykholt, near Thingvalla, has violet summits. On the coast, at the foot of the snowy mountains, extend black and red beaches of a blood-colour, according to the nature of the sand. The waters of the fiords, bordered with high basaltic rocks, have sometimes a reddish tinge also, proceeding from the decomposition of the rocks, or from the tint of the submarine streams transmitted through the prism of waters. Floating icebergs, carried by the polar currents, are often stranded in these fiords, and there spread the freshness of their beautiful blue tints, enhanced by the striking green of the sea which surrounds them.

The glaciers, like immense diamonds set in the lava, light up the dark valleys which surround the volcanoes, the black sides of which have bright and sinuous lines traced by the foam of the torrents. It is near one of these—the Silfurdœkir ('stream of silver')—that the largest deposit of spar in Iceland is found. This beautiful crystalline mass, so transparent and pure, white like mother-of-pearl, forms here a thick vein, in the middle of which the torrent has made its bed.

During the summer days, the purity of the air and the limpidity of the light add to the strange beauty of these contrasts of form and colour, thus producing magnificent pictures, and transforming Iceland into a fairy country. But in the dark days, which are so numerous, the aspect of this poor and poetical

Fig. 19.—March through the Lava.

land, situated between the frosts of the pole and the fire of the abyss, is quite different.

Lord Dufferin thus describes one of these dull landscapes: 'A heavy, low-hung, steel-coloured pall was stretched almost entirely across the heavens, except where along the flat horizon a broad stripe of opal atmosphere let the eye wander into space, in search of the pearly gateways of Paradise. On the other side rose the contorted lava mountains, their bleak heads knocking against the solid sky, and stained of an inky blackness, which changed into a still more lurid tint where the local reds struggled up through the shadow that lay brooding over the desolate scene. If within the domain of nature such another region is to be found, it can only be in the heart of those awful solitudes which science has unveiled to us amid the untrodden fastnesses of the lunar mountains.'

FORMATION OF ICELAND BY VOLCANIC ERUPTION.

We have already stated the opinion of Mr. Scrope, who considers that Iceland has been formed by eruptions from the same volcano. A careful observer, Von Troil, who visited the island in 1783, also attributed its formation to successive volcanic eruptions, the products of which have slowly accumulated. M. E. Robert believes there was first, as a basis, an archipelago of primordial rocks, but that the present geological relief is owing to a great number of eruptions. He believes also that these eruptions, during which enormous masses of ignited matter have been vomited, have caused a great hollow under the central part of Iceland, which will give way in a catastrophe similar to those observed in the Andes and in the isle of Java, where, as we shall see further on, gigantic volcanic domes have suddenly disappeared in the bowels of the earth.

Admitting such a sinking, the radial inclination

of the old rocks of Iceland round the centres of eruption may be explained—an inclination which is the cause of the singular aspect generally seen on the coast. All the mountains which belong to the old system of the island are also inclined towards the large volcanoes.

Fig. 20.—Ruins of Dverghamrar.

A considerable portion of the volcanic formation of Iceland consists of beds of basalt, which form colonnades, giving to it a very monumental aspect. The ruins of Dverghamrar are principally worth notice, where, on two sides of a large circle, are arranged rows of vertical columns, again covered by other smaller columns differently shaped.

The Icelanders use the natural barriers created by the lava to enclose their flocks. In some places, great quantities of small craters, which are from twenty to thirty feet high, and which are probably the result of the contact of lava with water, are used also for sheep-folds, by means of openings at the base, the top being scarcely pierced. Near Raudholar, some of these craters, entirely open, enclose gardens, which are thus sheltered from wind.

Caverns also have been formed by the lava, which, preserved by the solid crust already existing at the exterior of the strata, flowed in a sort of interior canal, keeping its fluidity, which enabled it to make its exit. The immense cavern of Surtschellir, proceeding from a similar stream, is one of the most remarkable of those large volcanic veins. 'Its walls are hung with lava-stalactites. Toward the middle of the canal, under a kind of dome, the visitor is arrested by a dazzling mass; it is snow, which has accumulated there, after having penetrated into this hollow, lighted with a mysterious light by means of a little opening which time has made in the roof of the vault. At the time when we visited this vast natural glacier, a ray of sunlight penetrated obliquely through this passage, lessening the light of the torches with which we were provided, and rendered the interior still more gloomy. Quite at the accessible extremity of the canal, which falls a little, we penetrated into a gallery of fairylike magnificence, everywhere hung with the purest crystal, which the light of our torches reflected in a thousand ways. The ceiling was covered with brilliant spangles; and to the right, on the side, we noticed a set of, as it were, organ-pipes, or very beautiful stalactites and stalagmites of ice.'*

In some cases, conical mountains, composed of a

Fig. 21.—Craters serving for Sheepfolds.

great number of beds of different kinds, have taken an exactly pyramidal form, resulting from a slow degradation of the volcanic materials. The mountain

Fig. 22.—Cavern of Surtschellir.

of Tungu-Kollur, where the snow rests on the step-like sides of the strata, resembles, in gigantic proportions, the pyramids of Egypt.

* *Voyage de la Recherche.*

The tremendous denudation that the soft soil of Iceland has been subjected to for thousands of years, the decomposition caused by the atmospheric agency of ashes and scoriæ, tend to level the roughnesses of the volcanic rocks, and to reproduce fertile lands and pasturages that have been annihilated by a long series of eruptions. The falling down of the high rocks which border the gulfs also accumulates *débris*, which time levels and transforms into dunes of sand. These dunes continue to spread, the action of the sea becoming less, probably in consequence of a slow raising of the coasts, a phenomenon already observed in Scandinavia, Spitzbergen, Lapland, and Kamtschatka.

LAVA-STREAMS.

The fiords of Iceland, similar to those of the granitic coast of Scotland and Norway, have been formed by great lava-streams which the action of subterranean forces has raised up and disintegrated. These immense crevasses, which elevate their vast strata to very great heights above the sea, present in consequence of a series of slips and disintegration of the rocks, the aspect of castellated walls, large pyramids, monuments and ruins. To the south of the island, basaltic columns, caverns, and the natural arches of Stapi, remind one of the most curious formations in Ireland, and the beautiful grotto of Antrim in the Orcades. From a bed of basalt situated at the base of Snælfells-Jökul, rises a mountain called Kambell, which resembles a large Gothic cathedral. Other rocks have the appearance of Cyclopean walls, circles, feudal towers, and sphinxes, which suggest beforehand to the navigator all that island contains of strange and wonderful.

In his remarkable work on Iceland, M. Krug de Nidda gives a picturesque description of the fiords: 'These gulfs, which are sometimes but a half a mile broad, are five or six miles long among the moun-

Fig. 23.—Natural arch in Ireland.

tains, by which they are surrounded on all sides, some of the peaks rising to a considerable height.

'The upper half of these gigantic walls, covered with eternal snows, remains hidden in thick clouds; there no trace of life is found, all is death and solitude; there is no man—nothing human in the midst of these masses heaped up by nature; no forests or trees, but naked rocks that are generally too steep to give a holding place to the most humble vegetation; no other noise but the echoed breaking of the sea, no other movement but that of torrents, fed by the snows, which stream down the sides of the rocks like silvered ribbons.'

'I shall add for myself,' says M. E. Robert, after having quoted this description, 'that towards midnight at the time of year when the sun is always above the horizon in the northern countries, and when the atmosphere is of a most perfect purity and calm, there reigns at the bottom of these same fiords an indefinable mysterious light, which I have never seen elsewhere but in Iceland: one might consider them as so many sanctuaries where Nature reposes.'

IV

VOLCANOES OF THE ATLANTIC.

ISLE OF JAN MAYEN—THE ESK AND THE BEERENBERG—VOLCANIC FORMATIONS OF THE NORTH SEA—VOLCANOES OF THE AZORES AND THE CANARIES—PEAK OF TENERIFFE—ERUPTIONS OF 1704 AND 1798—THE HESPERIDES—CAPE DE VERDE ISLANDS—ASCENSION AND ST. HELENA—SUBMARINE VOLCANIC REGION—VOLCANOES OF THE ANTILLES—ERUPTION OF MORNE-GAROU—SOLFATARA OF GUADELOUPE—INFLUENCE OF SEAS.

ISLE OF JAN MAYEN.—THE ESK AND BEERENBERG.

THE principal volcanoes of the Atlantic are situated in a direction nearly parallel to the coasts of Europe and Africa, as may be seen by tracing a line passing by the island of Jan Mayen, Iceland, Feroe, the western isles of Scotland, the north of Ireland, the Azores, Madeira, the Canaries, Cape de Verde Islands, Ascension, St. Helena and Tristan d'Acunha.

We shall but rapidly indicate the more remarkable phenomena relative to these volcanoes, which rise from the depths of the Atlantic, and being often visited by navigators are known by the many descriptions and by the eruptions that have recently occurred in some of them.

The coasts of Greenland present no active volcano,

Fig. 24.—PEAK OF JAN MAYEN.

the only noticeable object being the massive beds of basalt and other volcanic formations. The western part of Greenland has changed in level and continues to subside. The ruins of old edifices, now covered by the sea, prove this sinking.

The island of Jan Mayen is situated in 71° N. latitude in the prolongation of the volcanic chain of Iceland. It contains a volcano 1,500 feet high, discovered and visited by Scoresby in 1817 and called by him the Esk, after the vessel commanded by the gallant explorer. To the south-west of the Esk rises another volcano, also discovered in the same expedition, and which, since the month of April 1818, has shot out a large quantity of cinders, whilst puffs of smoke rose from the crater of the Esk. The Beerenberg, situated in the north-east part of the island, of which the summit, 6,648 feet high, cannot be reached, is probably an ancient volcano.

VOLCANIC FORMATIONS OF THE NORTH SEA.

We have already described the volcanoes of Iceland; we will now take a *coup d'œil* of the ancient region of volcanic action which belongs to the other islands of the North Sea. Most of these centres of eruption are surrounded with masses of cinders and scoriæ, broad streams of lava spread in basaltic platforms similar to those of the Feroe Isles, where the ancients placed Thule, which they supposed to be the limit of the earth.

In the centre of Scotland are seen a great number

of little volcanoes which appear to have produced but small quantities of cinders, whilst in the western part of the island enormous hills have been formed by the accumulation of thick beds of lava: to these the country owes its characteristic features. We may mention, on account of the tradition attached to it, the volcanic hill situated near Edinburgh, and known by the name of Arthur's Seat, in me-

Fig. 25.—Fingal's Cave at Staffa.

mory of the hero who was the support of the last Celtic races and the legendary king of Christian chivalry.

On the western coast of Scotland, the group of the Hebrides contains the basaltic island of Staffa, celebrated for the magnificent colonnade which forms Fingal's Cave. Ireland, also, has for some time been noticed for its immense and picturesque basaltic causeways; the most remarkable of which, 'The

Giant's Causeway,' often described, is situated near the sea, in the county of Antrim. Great sheets of lava, thick beds of ashes and solidified scoriæ, form, in the northern part of the British Isles, the highest

Fig. 26.—Interior of Fingal's Cave.

chains of hills. Besides we know that from one end to the other, England is nothing but an immense mineral bed produced by plutonic action, and that its prodigious mines of coal and iron have been the principal sources of its industrial power.

THE AZORES AND CANARIES.

The vast volcanic formations of the coasts of Spain and Portugal belong, like those that we have just described, to very ancient periods, and we will not, therefore, delay our route towards the regions of the Atlantic, where the volcanoes are still active.

The largest island of the Azores, San Miguel, is remarkable for a large number of cones of ashes, which form a central volcanic chain crossing the isle from east to west. One of the craters, from which have flowed trachyte and basalt in every direction, is fifteen miles in circumference. The appearance of the little island of Sabrina, a short distance from the coast, is the only volcanic phenomenon that San Miguel has presented since historic time. We shall speak further on of the eruptions which accomplished this upheaval.

The isles of Pico and San Jorge, belonging to the same archipelago, contain volcanoes, the last eruptions of which took place in 1718 and 1812.

In Madeira, everything indicates the long continuity of volcanic action; it is the principal island of the group, and appears to have been raised from the bottom of the Atlantic under the influence of this powerful action.

We arrive now at the Canaries, where we shall stop to describe the isle of Teneriffe, a large volcanic mountain, whose principal cone or peak is 4,921 yards above the level of the sea.

In the same archipelago, Palma and Canaria contain enormous craters, surrounded with peaked ramparts, showing successive beds of volcanic rocks and conglomerates. Fuertaventura and Lancerota are entirely volcanic, and filled with orifices, which give free passage to deluges of lava. At Lancerota, these

Fig. 27.—Grotto of Antrim.

orifices nearly all date from the eruptions which from 1730 to 1736 opened a great fissure through the whole length of the island.

PEAK OF TENERIFFE—ERUPTIONS OF 1704 AND 1798.

The large crater of the volcano of Teneriffe forms a vast oval, in the centre of which rises the peak and two other cones, called Chahorra and Montana Blanca. On the top of the peak, covered with snow during the greater part of the year, is the still smoking crater called Caldera (caldron), the walls of which are formed of abrupt rocks; on them are found, towards the orifice, beautiful needle-shaped crystals of sulphur. In 1704 the last memorable eruption of this volcano took place, and the destruction of the little town of Guarrachico, a ady devastated, in 1645, by a terrible inundation due to torrential rains.

'Guarrachico was a pleasant town, surrounded with fertile fields and rich vineyards; it had, besides, a very good and commodious port. During the night of May 5, 1704, a noise was heard underground like that of a storm, and the sea retreated. When the day broke, and rendered visible the phenomenon which frightened the unfortunate inhabitants, the peak was seen covered with a fearful red vapour. The air was on fire, a sulphurous smell suffocated the frightened animals, who uttered lamentable groans or plaintive bleatings. The waters were covered with a steam similar to that over the boiling springs; all at once the earth moved and opened; torrents of lava flowed from the crater of Teyde,* and rushed into the plains from the north-west. The town, half swallowed up in the clefts in the ground, half buried by the vomited lava, disappeared entirely. The sea, returning to its bed, inundated the ruins of the port which had sunk down; waves and heaps of cinders occupy the site of Guarrachico, and at the present time remains of horses are found among the fragments of lava. The inhabitants tried to save themselves by immediate flight, but most of them made futile attempts; some were swallowed up in the clefts which, in filling, buried them alive; others, suffocated by the sulphurous vapours, fell asphyxi-

* The name given to the peak by the inhabitants of the Canaries.

Fig. 28.—PEAK OF TENERIFFE.

ated in their attempts to escape. Many of these unfortunates, however, escaped with much peril, and seeing from far their homes in flames, flattered themselves with the hope that they had escaped death, when nearly all were crushed by a hail of enormous stones, the last effect of the fury of the eruption, which, after sending forth innumerable stones, quieted down, still threatening.' *

After this outburst, the Canaries for more than a century did not experience any fresh disaster produced by the subterranean fires. But in the night from the 8th to the 9th of June, 1798, a frightful noise was heard at Teneriffe, followed by severe shocks, which preceded a violent eruption of the Chahorra.

At Teneriffe, as in the Lipari Islands, Iceland, in the Andes, and other centres of eruption, rivulets of lava are seen, having the appearance of glass. These vitreous streams are often composed of obsidian or volcanic rock, the colours of which vary from black to green and red to yellow. Obsidian is worked into mirrors and objects of ornament. The Greeks carved it into points for arrows; the Mexicans employed it for the same use, and in making sword-blades. The first inhabitants of the Canaries, the Guanches, also manufactured instruments out of the volcanic materials, and tipped the extremity of their pikes, clubs, and javelins with them.

* *Les îles Fortunées.* By Bory de Saint-Vincent.

THE HESPERIDES.

The Canaries were known to the ancients under the name of the Fortunate Islands, which was doubtless given them by the first navigators on account of the beauty of their climate and the fertility of the soil. Beyond these islands 'one only saw the region where the day ended, and where the vault of the sky rested on the globe.' Poets made it the resting-place of happy souls, and in the Odyssey, the sea-god Proteus says to Menelaus, 'the immortal gods will send you to the Elysian Fields, to the extremity of the world, where the sage Rhadamanthus is the lawgiver, where men pass a happy and peaceful life, where there is neither snow nor the regions of winter, but where the air is always refreshed by the sweet breathings of the zephyrs sent by the ocean.' According to Hesiod, Jupiter placed the souls of heroes 'at the extremities of the world, beyond the deep gulfs of the ocean, in the Fortunate Islands, where thrice a year the fertile soil pours forth fruits brilliant and sweet as honey.'

Ancient poems also describe to us 'the enormous Atlas in the midst of the ocean, before the Hesperides, upholding the vast heavens. The snowy peak of which, covered with pines and beaten upon by the winds, is continually surrounded with obscure clouds, and illuminates the night with the fire which escapes from it.'

At Teneriffe, as at Madeira, and also on the slopes of Vesuvius and Etna, the vine grows by the side of the palm, and produces excellent wine. In MM. Barker-Webb and Berthelot's *Histoire naturelle des îles Canaries,* they thus describe a landscape which the traveller meets with in one of the parts of the island all planted with the vines of Malvoisie: 'Now the obstacles multiply; we travel through an ancient country; the ground is rugged, filled with holes and roughnesses; but plants grow vigorously in these fields, where formerly eruptions gave rise to flames. The fruits, more tasty than elsewhere, are there always early. We are now near the grotto of Icod, a gloomy cavern which undermines the whole valley. Nevertheless, the banks widen, the sea extends its horizon in the distance; we cross the wooden bridge that has been thrown across the ravine, and soon, in turning the buttresses of the Vega, Guarrachico reveals to us its burnt shores. The tide breaks against the steep rocks of Guincho; a torrent rushes from the top of the rocks, and falls in a noisy cascade at a few paces from the beach, near a group of bananas. Nothing can hinder the careful vinedresser; cultivation adorns all the cliffs which border the coast, and the green vine leaves cover the mountain from the base to the summit; but in the neighbourhood of the town, the arid, black, calcined rock forms a striking contrast; there is nothing now but confusion; a great disaster is revealed to us; we pass into streets filled with lava, and we can walk only among ruins.'

CAPE DE VERDE ISLANDS—ASCENSION—ST. HELENA.

Before continuing our excursion across the Atlantic, we must mention as a remarkable fact the scarcity of active volcanoes on the coast of Africa. The volcanic islands of Annobon, St. Thomas, Prince Island, and Fernando-Po, are nevertheless situated, according to Humboldt, on a line directed towards the chains of the Cameroon Mountains, in which an eruption of lava was observed in 1838, on the volcano Mongo-ma-Leba.

The Cape de Verde Isles appear to have quite a volcanic origin. The peak of Fuego, in activity at the present time, is 2,843 yards high; its last eruptions date from 1785 to 1799. Vast craters, high and steep rocks composed of basaltic beds, cones of scoriæ, streams of lava, prove the old volcanic action in the other islands in the archipelago. The peak of Fuego, like Stromboli, sent out flames without interruption from 1680 to 1713. Ascension and St. Helena, entirely volcanic, are covered with masses of lava and *débris* thrown out in the late eruptions. Among this *débris* in Ascension, around a vast open crater, a great quantity of volcanic bombs not less than six inches in diameter is seen; these bombs are produced by the rotatory movement imparted to some of the liquid lava thrown into the air by the craters. In the last eruption of the Kotlugaia in 1860, 'By night, a fountain of balls of fire (bombs) rose to a height calculated at 24,000 feet, since it was seen at sea 180 miles off. Many of these volcanic bombs, it is said, were seen and

Fig. 29.—VOLCANIC ROCKS OF ST. HELENA.

heard bursting in the air with a loud detonation at a distance of 100 miles. They must therefore have been of large size, as well as thrown up to a vast height. Their bursting there is rendered highly probable by the fact that fragments of volcanic bombs, evidently having this origin, are occasionally found near eruptive vents, and if we suppose the surface of a globular mass of liquid lava to be consolidated as it were with a rotating movement to a great height in the air, the expansion of the contained gases in the rarefied atmosphere in which it finds itself at its extreme height is very likely to cause the bursting of the shell with a loud explosion.'

According to Darwin (*Volcanic Islands*), Saint Helena is formed by a large circle, the basaltic ramparts of which trace round the island a belt of steep black rocks, their perpendicular height being from 100 to 2,000 feet. This circle is the last vestige of an enormous volcano, nearly entirely filled by the eruptions of a more modern one, the crater of which forms a precipice, now divided into peaks, in the central chain of the island. At Tristan d'Acunha, in the middle of a group of little volcanic islands, the crater of a volcano is observed, which has been elevated in the crater of a much more ancient volcano.

Among the volcanic islands situated in the South Atlantic Ocean, we may notice that of Deception, that was seen, in the month of February 1842, to vomit flames on thirteen different points situated in a circle; and the isles of Amsterdam and St. Paul, both of them containing active volcanoes.

SUBMARINE VOLCANIC REGION.

In a very interesting communication to the Academy of Sciences,* M. Daussy has been the first to call attention to the probable existence of a submarine volcano in a point of the Atlantic Ocean situated to the south of the Equator, where seamen have often observed strange phenomena: 'There are many examples of upheavals which caused islands to appear on the surface of the waters, the existence of which islands was only momentary. Such are the Isle of Julia in the Mediterranean, and those which appeared in the Azores in 1721 and 1811. The attentive examination of all the indications furnished by navigators has led me to believe that a similar phenomenon may have been produced some miles south of the

Fig. 30.—Isle of Julia.

* *Comptes rendus*, t. vi. 1838.

Equator, towards the twentieth or twenty-second degrees west longitude; or at least the shocks felt by different ships in those parts indicate the existence of a volcano disturbing the ground from time to time which overlies it. We know that earthquakes which are felt at sea produce on vessels a similar effect to a

Fig. 31.—Submarine Eruption observed in the Atlantic.

shock against some rocks or against the bottom. Thus, in that which took place in 1835 on the Chili coast, and which extended 15° from north to south, and 10° from east to west, the vessels under sail or at anchor felt the shocks as if they had touched some rocks. It is, then, probable that when a vessel undergoes a similar shock in a place where the depth renders it impossible that they could have touched, it may be attributed to the effect of an action of this kind. Now it happens that different notes of shocks more or less strong have been made in the vicinity of the point to which we have referred, which is mid-way between the western coast of Africa and the eastern coast of South America, in the part where they are nearest together—that is to say, between Cape Palmas and Cape St. Roque.'

VOLCANOES OF THE ANTILLES.—ERUPTION OF THE MORNE-GAROU.

From Davis's Straits as far as the Straits of Magellan, throughout the whole extent of the western coasts of the Atlantic, scarcely any traces of volcanic action have been found, if perhaps we except the archipelago of the Antilles, where violent eruptions, earthquakes, boiling springs, and solfataras prove the energy of the subterranean fires.

Among the still active volcanoes, that of the isle of St. Vincent, called the Morne-Garou, after remaining a long time as a simple solfatara, was the scene of two great eruptions in 1718 and in 1812. Humboldt adds, in connection with this last eruption, some striking coincidences: 'The first movements, near the crater, began in May 1811, three months after the island of Sabrina had risen from the sea in the Azores. In the mountain valley of Caracas, 3,496 feet above the sea, they began to be faintly felt in December of the same year. The complete destruction of the large town of Caracas took place on March 26 of the following year 1812. As the earthquake which destroyed Cumana on December 14, 1796, was with reason ascribed to the eruption of the volcano of Guadeloupe at the end of September 1796, so the destruction of Caracas appears to have been an effect of the reaction of a more southerly West-Indian volcano—that of the island of St. Vincent. The dreadful subterranean noise resembling thunder, or the sound of a heavy cannonade, caused by a violent eruption of the last-named volcano on April 30, 1812, was heard in the wide grassy plains (the Llanos) of Calabozo and on the banks of the Rio Apure, 192 geographical miles to the west of its confluence with the Orinoco.' (Humboldt, t. ii. p. 14.) 'The volcano of St. Vincent had not sent forth any lava between 1718 and 1812; on April 30 a stream of lava issued from the crater's summit, and in the course of four hours reached the sea-shore. It is a curious fact, and one that was confirmed to me by very intelligent persons engaged in coasting navigation, that the noise was heard with much greater intensity far out at sea than in the immediate vicinity of the island.' (*Cosmos.*) It must also be remarked that Rio Apure is situated 210 leagues from the volcano, that is to say, the distance of Vesuvius from Paris. We shall meet with other examples of this subterranean propagation of volcanic rumblings, which have often been heard at great distances, and which, under certain circumstances, indicate to us, as justly remarked by Arago, 'the existence of a general action of the interior mass of the globe against the solid shell, as now constituted.'

THE SOLFATARA OF GUADELOUPE.

The road which leads to the summit of the Solfatara is very difficult and strewn with calcined stones. The ground, red like ochre, resembles the residue of the distillation of vitriol. At a certain height, in a space about twenty-five fathoms in diameter, only sulphur, cinders, and carbonised earths are met with. Here several deep clefts open, whence escape vapours, sometimes mingled with flames, and at the bottom of which a boiling noise is heard. Sulphur also is given out, which attaches itself to the walls of these clefts, and the sulphurous acid, separated by the heat, condenses into drops, and runs out like clear water. The surface is scarcely solidified, and if one does not take precaution, there is a risk of being swallowed up.

This locality seems to be the opening whence former eruptions have proceeded. It is stated that this mountain was split in two during an earthquake, and sent forth a great quantity of flaming substances. In the plain to the north of this opening, which is 722 feet in diameter and 114 feet deep, is a little pond of which the water is strongly impregnated with alum.

The Solfatara of Guadeloupe produces sulphur of different kinds — some perfectly similar to flowers of sulphur; others in compact masses of beautiful yellowish gold; others still of a yellow as transparent as amber.

At St. Lucia, the condensation of the vapours given out by a crater named Oualibou forms also an active solfatara in the crater. This island contains moreover intermittent springs of boiling water, analogous to the geysers of Iceland. The islands of St. Eustache, Nevis, and Mont Serrat constantly give out sulphurous vapours, which indicate, like the mud volcanoes of Trinity Island, the activity of the volcanic forces.

INFLUENCE OF SEAS.

Volcanoes are almost always situated in the neighbourhood of the sea or of great lakes. It must also be remarked that volcanic phenomena have ceased suddenly when the latter have from any cause disappeared. This was remarked in Auvergne, in the case of an old lake named Limagne, which has now become one of the most fertile districts, rarely troubled by earthquakes. Is it not also from this cause that, in the larger volcanic islands like Iceland, volcanic phenomena are more common close to the sea than in the interior—nearer the great lakes than elsewhere?

The upheaval of the primeval mountains had, perhaps, no other cause than the enormous pressure of the sea when the waters, in the first ages of the world, precipitated by the condensation of the atmosphere, covered two-thirds of the globe. In the same ratio that the ocean augmented in volume did the depression of the basin which received it become greater; consequently, the central incandescent matter was driven back in several points; and to restore the equilibrium of nature, in which everything tends to a state of rest, this matter was compelled to reconstruct the shores of the ocean, first, in elevating the primitive mountains, and then in producing the numerous volcanoes which seem to continually diminish. We must also, with Arago, recognise the fact that the bottom of the sea and the shores situated far below the continental levels would present smaller resistance to the subterranean forces, and thus more easily permit eruptions to take place.

The volcanoes which are still active in Central Asia, which are seen in the middle of the chain of Thian-Schan, or the Celestial Mountains, are, it is true, far removed from the Indian and Frozen Oceans; but these volcanoes, as remarked by Humboldt, are situated very near the great depression which formerly formed a great basin, since divided into a series of lakes, called by the French *Lacs à chapelet.* Old traditions speak of a dried-up sea, and Humboldt quotes in this connection a curious fact:—

'Some sea-calves, precisely like those which congregate round the Caspian Sea and the Baitral, were found at a distance of 100 geographical miles, in the little lake of Oron, which is filled with fresh water and is but a few miles in circumference, whilst there exist none in the Lena, although the river Witim, one of its affluents, is in communication with Lake Oron. The isolation in which these animals now live, the distance which separates them from the mouth of the Volga, a distance equal to 900 geographical miles, is a remarkable geological fact, which indicates a vast and ancient system of communication between the waters. Could the immense and numerous depressions which the surface of Central Asia has been subjected to have had exceptionally the same influence on the continental upheaval, and created the same relations which are produced near shore, in the neighbourhood of upheaved beds, by the depression of the sea bottom?'

V

THE ANDES.

CHAINS OF VOLCANOES—CRATER OF PICHINCHA—ERUPTION OF COTOPAXI—FALLING IN OF CARGUAIRAZO—ERUPTION OF SANGAY—THE ANTISANA—VOLCANOES OF CHILI AND PERU—THE HELL OF MASAYA—ERUPTION OF COSEGUINA—THE IZALCO—AGUA AND FUEGO—VOLCANOES OF MEXICO—THE MALPAIS—THE COMPANIONS OF CORTEZ ON POPOCATEPETL—CRATER OF ORIZABA—OUTBREAK OF JORULLO—NORTHERN VOLCANOES.

CHAINS OF VOLCANOES.

THE traveller who follows the western coast of America from Cape Horn to Behring's Straits, frequently sees volcanic heights rise from the midst of the majestic mountains which dominate the Pacific Ocean on one side, and the basins of the large rivers on the other. Not less than 115 craters may be counted, by which the interior forces of the globe communicate with the atmosphere. All the energy of the subterranean forces seem to be concentrated in the New Continent on a single line, for no other trace is found of them.

These forces, moreover, surpass many of those the effects of which we have already described. The giants of the Andes are nearly double the height of Etna. Generally they vomit only scoriæ, ashes, and smoke, but sometimes lava flows from their craters, as seen on the Antisana. The summit of this volcano is 6,379 yards above the sea level. If we calculate the pressure necessary to sustain a column of lava at this height, we find one of 1,500 atmospheres is required, whilst one of 300 only is sufficient to raise the lava to the mouth of Etna. We shall be better able to form an idea of the magnitude of these forces, if we remember that a pressure of one atmosphere supports a column of water 32 feet in height, and that our most powerful steam-engines only work up to about ten atmospheres.

Leopold von Buch, who has given a great impulse to geology and particularly to the study of volcanoes, has established an important division. He has distinguished, in the first place, the class of central volcanoes which form groups in the middle of which lies the principal vent. Such are the volcanoes described in the preceding chapters. The second class is constituted by volcanic chains or volcanoes in lines, which are met with one after the other, as if they were the ventilating shafts of the same gallery. We find these especially in the Andes and in the Pacific Ocean. In the middle of the mountains of western America rectilinear ranges, which comprise as many as twenty volcanoes, extend at different points over a space of about 130 geographical miles (the distance from Vesuvius to Prague), and this chain is sometimes parallel to the general axis of the Andes and sometimes transversal. If we except the interesting observations of Bouguer and La Condamine, it may be said that this field of important research has not been explored since the end of the last century, the date of the voyages and discoveries of Humboldt and Bonpland, who are now about to serve us as guides. The volcanoes of Colombia are the most noted among those of the New World, and this celebrity is due to the scientific work of Bouguer and La Condamine in connection with them.

In the region of Quito are eighteen volcanoes, ten of which are still in eruption. The view presented by their high peaks, which are distributed in a picturesque manner, is especially grand in character when we consider the connection of every part of the chain by subterranean communication. A general focus seems to spread itself under the entire plateau, and it is observed that the centre of activity for centuries has slowly propagated itself in the direction from south to north.

THE CRATER OF PICHINCHA.

Quito is built at the foot of this volcanic mountain which during the sixteenth century was the scene of violent eruptions, and although quieter since 1660 is not yet extinct. It is formed by an immense wall-like mass of black trachyte, which follows the line of fault in the Cordilleras for a distance of nine miles. This wall, like many strong castles, has three domes, the principal of which, called Rucu-Pichin-cha (the Father or the Ancient One), reaches the region of eternal snows.

Humboldt, guided by an Indian, in 1802, climbed up the most eastern of these lofty peaks. Arrived, by very dangerous paths, to the extreme edge of the crater, he found himself at about 874 yards from the bottom of the fiery abyss. 'It was,' said he, 'a magnificent spectacle. Nature has never been pre-

Fig. 32.—The Pichincha.

sented to me under a grander aspect.' He afterwards mentions the observations made since that time in the very crater of the volcano by a learned traveller, M. Wysse, who did not fear remaining there several nights.

'This crater is divided into two portions by a ridge of rock covered with vitrified scoriæ. The eastern part, of circular form, and deeper than the other by 1,000 feet, is actually the real seat of the volcanic activity. It contains a cone of eruption about 250 feet high and surrounded by 70 still-burning openings, or 'fumaroles,' whence proceed sulphurous vapours. It is probably from the crater, covered at the coolest places with tufts of a kind of grass similar to reeds, that the igneous eruptions of scoriæ, pumice, and ashes, which succeeded each other in 1539, 1560, 1566, 1577, 1580, and 1660, proceeded. During these eruptions the town of Quito was often plunged for a whole day in complete darkness, caused by the dust of the rapilli.' *

ERUPTIONS OF COTOPAXI.—FALLING-IN OF CARGUAIRAZO.

The top of the volcano of Cotopaxi forms a perfectly regular cone, on which the snow-line and the limit of forest vegetation are marked in the clearest manner. Lower down, a few rough peaks, called 'the Head of the Inca,' seem to have been ancient streams of lava.

The approach of an eruption is announced by the sudden melting of the snows which cover the summit. Before the smoke rises in the air, the walls of the cone become incandescent and shine in the middle of the night with a reddish light, above the enormous black mass of the mountain.

In 1741 commenced a long and disastrous eruption; according to La Condamine, the columns of fire which rose from the volcano attained a height

* *Cosmos.*

of 1,000 feet. At the same time torrents of water, proceeding from the sudden melting of the snows which had been collecting for two centuries, rushed down the sides, drawing with them blocks of ice and smoking scoriæ; their power was so great that in the plain were seen large waves, and the

Fig. 33.—Craters of the Pichincha. (Humboldt.)

velocity of the waters, even at four leagues from the mountain, was, according to the estimation of Bouguer, 56 feet per second; 600 houses were destroyed, and nearly 1,000 persons perished.

Immense floods have sometimes been caused by the slow but continuous action of the snow during periods of repose. In consequence of its fusion, incessant infiltrations of water penetrate into the rocks. 'The caverns in the sides of the mountain or at its base are thus transformed by degrees into reservoirs of subterranean waters, put into communication with the alpine streams of the plateau of Quito by narrow channels. The fish of the streams prefer to spawn in the shades of the caverns, and when the shocks which always precede the eruption of the Cordilleras make the entire mass of the volcano tremble, the subterranean vaults fall in, and the water, fish, and tufaceous mud are instantly expelled. This singular phenomenon brought to the knowledge of the inhabitants of Quito the little fish which they call *prenadilla*. In the night from the 19th to the 20th of June 1698, the summit of the volcano of Carguairazo, 6,561 yards high, suddenly fell in, with the exception of two enormous pillars, the last vestiges of the ancient crater; the surrounding lands were covered, over an extent of nearly seven square leagues, with the moistened sandy stone and a clayey mud containing dead fish, and reduced to sterility. The pernicious fevers which broke out seven years later in the town of Ibarra, to the north of Quito, were attributed to the putrefaction of a great number of dead fish ejected from the volcano of Imbabaru.' *

It is to the great eruption of Cotopaxi, which took place in 1533, the terrible remembrance of which is preserved in the country from generation to generation, that the existence, at a distance of three leagues, of blocks of trachyte measuring 130 cubic

* *Cosmos.*

Fig. 34.- ERUPTION OF COTOPAXI (1741).

yards in volume is attributed. A circumstance which dispels all doubt as to the origin of these stones, is that they are arranged on every side in lines directed towards the volcano.

ERUPTION OF SANGAY.—THE ANTISANA.

On the east slope of the eastern Cordilleras, between two systems of affluents which increase the waters of the Amazons, is the large volcano of Sangay, 6,561 yards high, and nevertheless more active than the little cone of Stromboli, which only rises 984 yards above the sea. The eruptions of Sangay, which began in 1728, did service as a perpetual fire-signal whilst the French *savants* were engaged in measuring an arc of the meridian. These eruptions were accompanied with terrible noises, called by the inhabitants *bramidos*, which are sometimes heard at great distances. Thus, in 1842 and 1843, at the time when the subterranean thunders were more violent than ever, they were distinctly heard as far as Payta, along the coasts of the Pacific.

According to M. Wysse, who was the first to climb this colossal mountain, there are 267 eruptions in an hour; each lasts, on an average, thirteen seconds. The intrepid traveller proved also this remarkable circumstance, that, even on the cone of ashes, no perceptible shock accompanied these frequent emissions. The substances thrown out in the midst of a thick smoke, of a colour sometimes grey, sometimes orange-coloured, are ashes and scoriæ.

M. Wysse counted, in a great explosion, sixty scoriæ having a spherical form, and about two feet in diameter. Most of them again fall into the crater, or slide over the wall of the cone, throwing out a light such as La Condamine believed he saw in flames produced by sulphur and asphalte. The ashes, which were very black, gave a frightful aspect to the upper part of the volcano. On its sides, and for eleven miles round it, they were spread in beds which in some places were 400 feet in thickness.

Antisana is a volcano, the most recent eruptions of which took place in 1590 and 1718; it, however, deserves a particular notice. In the Cordilleras of Quito it is the only one which presents lava-streams.

In its contour it is like Sangay. At a height of 4,593 yards is an oval plain, whence rises, like an island, the part of the volcano covered with perpetual snows. The dome-like top is connected by jagged hills to a truncated cone situated to the north. Formerly the plain served as a bed to a lake, but now the sheet of water is reduced to a lagune. Ramparts of basaltic stones rise at the foot of the mountain. Many of these rocks are so scorified that they resemble sponges.

VOLCANOES OF CHILI AND PERU.—LAST ERUPTIONS. THE LAND OF FIRE.

Among the thirty-eight craters, spread in two groups on the Cordilleras, which extend from Columbia to the Straits of Magellan, sixteen may still be considered as active. The highest of all these volcanoes is the Sahama, which attains a height of 7,655 yards—six times that of Vesuvius. It is situated at the point where the chain of the Andes changes its course. It has a fine truncated cone, perfectly regular, covered with dazzling snow, and always crowned with a plume of smoke.

Near Arequipa are six volcanoes, one only of which is in action. In the sixteenth century this town was almost entirely buried by an eruption of ashes from Uvinas, now extinct and situated at a distance of several leagues.

The volcano of Gualatieri, in Bolivia, is still active, according to Pentland, who found an ancient crater with streams of lava at the foot of the eastern chain, more than forty-five geographical miles from the coast—that is to say, at a much greater distance than that of Sangay.

The disturbances of the earth's crust in these countries have brought to the surface the most precious mineral treasures, formerly consecrated by the Incas to ornament their temples of the Sun. On an island in the middle of the Lake Titicaca is the most ancient of these sanctuaries. It was an edifice covered with plates of gold, and it contained the famous chain of the same metal, about 700 feet long, which the Inca Huayna-Capac manufactured, and which was used in religious ceremonies. The Indians could only rescue this chain from the avidity of their conquerors by throwing it into the depths of the lake, which perhaps now preserves it. Let us mention also the celebrated mines of Potosi, worked in the Andes, in a region of porphyry rocks, at a height greater than that of Mont Blanc.

A remarkable particularity which is still seen in Peru consists of enormous beds of trachyte, which have been ejected, not by the craters of volcanoes but by lateral fissures. One has covered the ground over the surface of more than 200 square miles.

To reach the volcanic chain of Chili, we must pass over a space of 135 miles, in which distance not one is met with. A group of thirteen active craters, dominated by the Aconcagua, which rivals the Sahama in grandeur and beauty, rises above the coast. When a shock of an earthquake, so often felt in this country, takes place, long jets of flame and smoke are seen at the same instant to come from most of the volcanoes.

The Antuco vomits every quarter of an hour sulphurous vapours, ashes, and pumice. Its loud detonations are heard at a distance of twelve leagues. We shall revert again to the volcanic phenomena of Peru and Chili when we speak of the terrible earthquakes to which these countries are so often subjected.

The nature of the Patagonian Andes is very little

known. At the extremity of the American continent is the Land of Fire, or Terra del Fuego, which conceals a powerful volcanic focus under its thick mantle of snow.

On many parts of its coasts are basalts and porphyritic lava, with scoriæ. In the central region rises the cone of Sarmiento, the burning summit of which rises 2,405 yards above the level of the sea. The line of subterranean activity that we have just followed is continued in the southern seas by the southern Shetland Isles, in which many thermal springs are seen in the midst of the snow. Deception Isle is nothing more than a vast annular crater, the perpendicular rocks of which are composed of alternate beds of ice and lava. In the year 1842 it was seen to vomit flames on thirteen different points, arranged in a circle.

THE HELL OF MASAYA.

The volcano of Masaya, or, as it has been called, *El Infierno* (or Hell) *de Masaya*, which, at the commencement of the sixteenth century, had a very wide-spread reputation, and which has been the subject of memoirs addressed to Charles V., is situated in Central America, between the two lakes Nicaragua and Managua, to the south-west of the charming Indian village of Nindiri. During many centuries it has presented the rare phenomenon observed on Stromboli. From the sides of the crater are seen, through a fiery opening, waves of lava agitated by vapours. The Spanish historian, Gonzalès Fernando de Oviedo, who, in 1501, visited Vesuvius with the Queen of Naples, was the first to climb Masaya, in the month of July 1529, and made comparisons between the two volcanoes. The word Masaya signifies 'fiery mountain.' The volcano, surrounded by a vast field of lava, which it has doubtless formed, was considered at that time as belonging to the volcanic group of the Maribios. 'In the ordinary state,' says Oviedo, 'the surface of the lava, in the middle of which swim black scoriæ, remains several hundred feet below the edge of the crater; but sometimes it suddenly produces a boiling, so much so that the lava nearly attains the highest border.' The perpetual illumination of the Masaya proceeds, according to the ingenious and precise language of Oviedo, not from a flame, properly speaking, but from vapours lighted up from the bottom. This phenomenon was of such an intensity that on the road, nearly three leagues in length, which goes from the volcano to the town of Grenada, the country was nearly as light as at the time of full moon.

ERUPTION OF COSEGUINA.—IZALCO.—AGUA AND FUEGO.

Central America, in which in a chain of twenty-nine volcanoes there are eighteen still burning, forms, with the Sea of the Antilles, one of the regions of the globe where subterranean activity is most intense. We have described in the last chapter the phenomena which the interior portion presents; it remains for us to give some details of the volcanoes scattered along the Cordilleras, which rise between the two seas, over an extent of about six degrees of latitude.

In ascending Nicaragua, towards the north, to the Gulf of Fonseca, we encounter, at a distance of five miles from the Pacific, six volcanoes ranged in file and placed close together. They have the collective name of Los Maribios. The Coseguina, on the promontory which forms the southern extremity of the gulf, owes its celebrity to the terrible eruption which took place in the month of January 1835, the deep obscurity caused by the ashes lasting two days. Terrible detonations were heard in the peninsula of Yucatan; on the shore of Jamaica; and also on the plateau of Bogota—nearly 3,280 yards above the sea, and at a distance of 140 geographical miles. By a remarkable coincidence, the volcanoes of Aconcagua and Corcovado, in Chili, began eruption on the same day.

The eruptions of the volcano of Izalco, situated in a plain to the east of the port of Sonconate, take place four times an hour, and serve as a lighthouse to mariners who are overtaken by night in these parts. Its cone, which is 1,640 feet high, was suddenly upheaved like the volcano of Jorullo, of which we will give a description further on.

Two volcanoes of the same chain have the strangely contrasted names Agua (water) and De Fuego (fire). The first owes its title to the sudden melting of the snows at its summit, which occasioned a disastrous inundation of the neighbouring town of Guatemala.

The second was the scene, in the last century especially, of great eruptions, accompanied with earthquakes; and again very recently it sent out an enormous torrent of lava. The government of the province should force the inhabitants of a town subjected to so many disasters to settle themselves in a more northern site.

VOLCANOES OF MEXICO.—THE MALPAIS.

From the Soconusko, which terminates the chain of Central America, to the totally different system which characterises Mexico, no volcanic formation is met with in a course of forty miles. All the peaks which compose this system appear laid out by line, as if they had proceeded from a single crevasse, its length about ninety miles, in a perpendicular direction to that of the great chain of mountains which crosses Mexico from north to south. This parallel of volcanoes, as Humboldt calls it, oscillates but a few minutes about the geographical parallel of 19°. It has been remarked, moreover, that prolonging this line 110 miles to the west of the coasts of the Pacific Ocean it meets the isles of Revillagigedo, in the vicinity of which pumice-stones often float in great quantities, and that further on it would lead to the great volcano of the Mauna-Roa, in one of the Sandwich Islands. Similar correspondences probably indicate traces of dislocation of the earth's crust, conformably to a geological theory of which we shall speak presently.

Of the six Mexican volcanoes—Orizaba, Toluca,

Tuxtla, Popocatepetl, Jorullo, and Colima—the four last are still in activity, or have been burning within historic times. On the plateau whence they rise, in many localites are found, on the surface of the ground, vast fields of lava, entirely deserted, to which the inhabitants give the significant name of *Malpais*, and which testifies the extreme energy of the subterranean forces.

A similar denuded bed is situated to the west of Puebla, at the foot of the volcano of Popocatepetl. It is three miles long, one mile broad, and rises from 65 to 100 feet above the neighbouring plain. Blocks of black lava, sometimes upright, and scattered here and there with spots of yellowish pumice, and some scanty lichen, present a very savage aspect. These enormous masses do not appear to be the result of lateral overflowings. 'It is probable,' says Humboldt, 'that at the time of the upheaval of the mountain, the bending of the ground produced over a vast space longitudinal openings, and a network of fissures, whence have risen substances in a state of fusion, sometimes under the form of compact masses, sometimes as scorified lava, without forming the scaffoldings of mountains, that is to say, open cones or craters of upheaval.'

THE COMPANIONS OF CORTEZ ON THE POPOCATEPETL.

We read in the accounts of Cortez that, having been very much struck by the sight of Popocatepetl always burning, he sent his courageous companions to the summit 'to discover the secret of the smoke,' a secret which he wished to impart to Charles V. This episode is thus recounted by the historian, W. Prescott:*

'They were passing between two of the highest mountains on the North American continent, Popocatepetl, "the hill that smokes," and Iztaccihuatl, or "white woman"—a name suggested, doubtless, by the white robe of snow spread over its broad and broken surface. A puerile superstition of the Indians regarded these celebrated mountains as gods, and Iztaccihuatl as the wife of her more formidable neighbour. A tradition of a higher character described the northern volcano as the abode of the departed spirits of wicked rulers, whose fiery agonies in their prison-house caused the fearful bellowings and convulsions in times of eruption. It was the classic fable of antiquity. These superstitious legends had invested the mountain with a mysterious horror which made the natives shrink from attempting its ascent, which indeed was, from natural causes, a work of incredible difficulty.

'The great volcano, as Popocatepetl was called, rose to the enormous height of 17,852 feet above the level of the sea; more than 2,000 feet above the "monarch of mountains," the highest elevation in Europe. During the present century, it has rarely given evidence of its volcanic origin, and "the hill that smokes" has almost forfeited its claim to the appellation. But at the time of the conquest, it was frequently in a state of activity, and raged with

* *History of the Conquest of Mexico*, book iii.

Fig. 35.—CRATER OF POPOCATEPETL.

uncommon fury while the Spaniards were at Tlascala —an evil omen, it was thought, for the natives of Anahuac. Its head, gathered into a regular cone by the deposits of successive eruptions, wore the usual form of volcanic mountains when not disturbed by the falling-in of the crater. Soaring towards the skies, with its silver sheet of everlasting snow, it was seen far and wide over the broad plains of Mexico and Puebla, the first object which the morning sun greeted in his rising, the last where his evening rays were seen to linger—shedding a glorious effulgence over its head, that contrasted strikingly with the ruinous waste of sand and lava immediately below, and the deep fringe of funereal pines that shrouded its base.

'The mysterious terrors which hung over the spot, and the wild love of adventure, made some of the Spanish cavaliers desirous to attempt the ascent, which the natives declared no man could accomplish and live. Cortez encouraged them in the enterprise, willing to show the Indians that no achievement was above the dauntless daring of his followers. One of his captains, accordingly, Diego Ordaz, with nine Spaniards, and several Tlascalans, encouraged by their example, undertook the ascent. It was attended with more difficulty than had been anticipated.

'The lower region was clothed with a dense forest, so thickly matted that in some places it was scarcely possible to penetrate it. It grew thinner, however, as they advanced, dwindling by degrees into a straggling, stunted vegetation, till, at the height of somewhat more than 13,000 feet, it faded away altogether. The Indians, who had held on thus far, intimidated by the strange subterraneous sounds of the volcano, even then in a state of combustion, now left them. The track now opened on a black surface of glazed volcanic sand and of lava, the broken fragments of which, arrested in its boiling progress in a thousand fantastic forms, opposed continual impediments to their advance. Amidst these, one huge rock, the Pico del Fraile, a conspicuous object from below, rose to a height of 150 feet, compelling them to take a wide circuit. They soon came to the limits of perpetual snow, where new difficulties presented themselves, as the treacherous ice gave an imperfect footing, and a false step might precipitate them into the frozen chasms that yawned around. To increase their distress, respiration in these aërial regions became so difficult, that every effort was attended with sharp pains in the head and limbs. Still they pressed on, till, drawing near the crater, such volumes of smoke, sparks, and cinders were belched forth from its burning entrails, and driven down the sides of the mountain, as nearly suffocated and blinded them. It was too much even for their hardy frames to endure, and, however reluctantly, they were compelled to abandon the attempt on the eve of its completion. They brought back some huge icicles, a curious sight in these tropical regions, as a trophy of their achievement, which, however imperfect, was sufficient to strike the minds of the natives with wonder, by showing that with the Spaniards the most appalling and mysterious perils were only pastimes. The undertaking was eminently characteristic of the bold spirit of the cavalier of that day, who, not content with the dangers that lay in his path, seemed to court them from the mere quixotic love of adventure. A report of the affair was transmitted to the Emperor Charles V., and the family of Ordaz was allowed to commemorate the exploit by assuming a burning mountain on their escutcheon.'

The conquerors of science have completed the enterprise of the companions of Cortez. Several naturalists have reached the summit of Popocatepetl, and have even begun to work the sulphur deposited there. We give, after a drawing of M. Jules Laverrière, a view of the crater taken from a breach at the north-east; it is a vast circular basin, the vertical walls of which are composed of strata of a pale rose colour in some parts, and black in others. Two snowy peaks dominate it. A capstan is used to descend to the interior platform. There many orifices are found, to which the name of *respiradores* is given; from them escape great columns of vapour—first red, then yellow, and lastly white. Their number varies, as a heap of fragments is sufficient to stop a funnel, and to turn or divide the ascending current. In 1857 there were five, the greatest diameter being about 19 feet. On their edges is sulphur seen in compact masses, which break with a bright surface of great purity; in granules mixed with sand; or in the state called 'flowers of sulphur.' In each year 800 metric quintals are worked.

CRATER OF ORIZABA.

Orizaba, a beautiful cone 6,561 yards high—the Aztec name Citlalpetl signifies 'mountain of the stars'—has been the scene of very violent eruptions, from 1545 to 1560, but since that time it has remained in repose. When Baron Müller ascended it in the month of September, 1856, he reached the top only in his second attempt. Lost in the first in the midst of glaciers, he was overtaken by a frightful tempest, and reached the bottom with much difficulty by an extremely dangerous route. Fresh information showed him a better road; two inhabitants of a neighbouring town joined him, and some Indians were sent forward to a grotto near the limit of the snows, to prepare that which was necessary to pass the first night.

The travellers did not reach this halting-place till long after the setting of the sun, when already the moon had cast her serene light on the vast and magnificent country. They contemplated the place after having warmed themselves by the fire lighted by the Indians. On one side, a curtain of black firs stood out from the sky; from the other, the gigantic volcano, nearly veiled by the mist, reflected the rays of the moon, and this mysterious light made it appear still more majestic.

The next day, to get beyond the fields of snow, it was necessary to help themselves on by crawling along the rocks with which they are covered. The difficulty of arriving at the top increased more when a fine snow began to fall; and it was not until six in the afternoon that M. de Müller with his companions could reach the edge of the crater.

'I had attained my end,' says he in his account, 'and joy caused all my griefs to vanish, but it was only for an instant, and a stream of blood flowed from my mouth.

Fig. 36.—Crater of Orizaba.

'When I came to myself, I was still near the crater; then I gathered all my strength to look and observe as much as possible. My pen cannot describe the aspect of these places, neither the impression which they produced on me. It is the door of the infernal world, where night and terror always reign. What terrible power there must have been to raise and burst these enormous masses, to smelt them and heap them up like towers, until the moment when they were cooled, and attained their present forms!

'A yellowish bed of sulphur in many places covered the interior walls, and at the bottom rise various little volcanic cones. The surface of the crater, as far as I could see, was covered with snow, and by no means warm in consequence. The Indians assured me that at different places warm air comes out from the clefts in the rock. Although I did not prove this, the fact appeared quite admissible, as I have often observed a similar phenomenon on the Popocatepetl.

'My first plan, to pass the night at the crater, had become impracticable. The twilight, which in this latitude is very short, had already set in; we were obliged to think of returning. The two Indians rolled the straw mats together which they had brought, and rolled them up in front in such a way as to form a kind of sledge; we sat on it, and, extending our legs, we slid down on this vehicle. The rapidity with which we were precipitated increased so much that our descent more resembled a fall in the air than any other means of locomotion; in a few minutes we cleared a space which had taken us five hours to climb.'

APPEARANCE OF JORULLO.

On the great line of fault near the parallel of 19° which we have noticed, between the volcanoes Toluca and Colima, but at a distance of 93 miles

Fig. 37.—JORULLO.

from each of them, suddenly appeared, in 1759, a new volcano, 1,640 feet high. This nearly contemporary phenomenon calls to mind the revolutions of the primitive periods of our planet, and is of the greatest interest to science. Humboldt made a complete study of it, in which he united to the observation of the places all the most important facts which tradition had preserved in the country on this terrible catastrophe.

Let us transport ourselves to the western slope of the Mexican plateau, where the vast plains of the province of Mechuacan extend, which enjoy a temperate climate on account of their elevation of 874 yards above the sea, and are renowned for their beautiful plantations. Between two small streams of water, called Cuitimba and San Pedro, there were, until the middle of the last century, fields cultivated with cotton, sugar-canes, and indigo, one of the richest *haciendas*, or rural properties, of the country.

From June 29, 1729, frightful subterraneous noises and numerous shocks of earthquakes succeeded each other for two months, plunging the inhabitants into consternation. Calm appeared to return at the beginning of September, but soon sinister signs appeared afresh.

On the 28th a phenomenon was observed which, as usual, rather marks the end than the beginning of the eruptions. Some labourers had gone out to gather guavas, which grew where the Jorullo now rises. When they returned to the *hacienda*, it was remarked with surprise that their hats were covered with volcanic ashes. Crevasses had already opened in the neighbourhood. At the same time the subterranean shocks became more and more violent, and in the first hours of the night the ashes were a foot deep.

'Everyone,' says Humboldt, 'fled to the heights of Aguasarco, an Indian village 2,400 feet above the old plain of Jorullo. From these heights they saw—so says the traditional account—dreadful outbursts of flame over a wide extent of country; and in the midst of the flames—according to the expression of the actual spectators of the upheaval of the mountain—there appeared a large shapeless lump, like a black castle. In the then very sparsely-inhabited state of the country, not a single human being lost his life in the long-continued earthquake, whilst at the copper-mines of Inguaran, in the little town of Patacuaro, houses were overthrown.' The area devastated comprised more than three square miles. To the pieces of rock, scoriæ, and ashes sent into the air were added an emission of muddy water, and enormous jets of vapour. The volcano rose nearly in the middle of the country, thus transformed into a *mal-pais.* It was composed of six cones of different sizes, the highest of which bears the name of Jorullo. The most important phenomenon which accompanied this appearance is the upheaval of a circular surface having a radius of 6,000 feet. This surface was surrounded with an escarpment 40 feet in height, of convex form, and in the centre it is raised 524 feet above the exterior plain.

Two rivulets, by which the country was watered,

disappeared in a deep crevasse of the eastern side. They now, doubtless, flow in volcanic subterranean passages, as they reappear at the west at a distant point from their ancient bed, forming two cascades, their waters being of a high temperature.

To give a better account of the state of the vast upheaved dome, we must have recourse to a report in which the commissary of mines (Fischer) relates descriptions of it by eyewitnesses. 'Before the dreadful mountain appeared, the earthquake shocks and subterranean noises increased more and more; that on the day itself the flat ground was seen to rise visibly, and the whole more or less swelled up, so that blisters appeared, of which the largest is now the volcano. These raised blisters, of very different magnitudes, and some of them of tolerably regular conical form, afterwards burst and emitted boiling-hot mud, as well as scoriaceous rocks, which are still found, covered with black masses of stone, to an enormous distance.'

The many thousand small cones of eruption, which are pretty equably distributed over the Malpais, are generally from four to about nine feet high: the smoke escapes by lateral openings, and not by the top. Hence the name *hornitos* (little ovens). When we approached them, and listened attentively, we heard a sound like the fall or rush of water, the contact of which with the incandescent masses occasions the formation of columns of vapour, which often spread in foggy banks, behind which appeared the dark masses of the volcanic hills.

'We may in some degree infer,' remarks Humboldt, 'the heat which must have prevailed in the surrounding atmosphere from the temperatures still measured by me after a lapse of forty-three years; we are reminded thereby of the early state of our planet, when under every latitude, and for long periods of time, the temperature of the atmosphere and the distribution of organic life were subject to modifications arising from the thermic influence of the interior acting through deep fissures.'

The six cones of the group of the Jorullo are distributed along a fissure nearly a mile long, and divided in such a way as to cut at a right angle the line formed from one sea to another by the great volcanoes of Mexico. These cones were in full eruption for about a year, but afterwards their activity rapidly diminished. Now there is scarcely any emission of vapour from their craters, most of which are filled with scoriæ.

NORTHERN VOLCANOES.

We shall only take a rapid glance on the north-west region of America, the volcanoes of which are yet little known. There does not exist, from Mexico to the extremity of the Rocky Mountains, uninterrupted chains, but an immense swelling of the ground, which, increasing in breadth, lengthens in the direction from north and north-west, thus continuing the line of the Andes by a vast plateau, on which groups of isolated mountains appear from distance to distance. These mountains are most frequently cones of trachyte, from 3,280 yards to 4,374 yards high, and they impress the traveller much more as the plateau seems to be a never-ending plain.

'On the eastern slope of the Rocky Mountains, on the south-western route from Bent's Fort on the Arkansas river to Santa Fé del Nuevo Mexico, there are two extinct volcanoes—the Raton Mountains with Fisher's Peak, and (between Galisteo and Pena Blanca) the hill of El Cerrito. The lavas of the first of these covered over the whole country between the Upper Arkansas and the Canadian river.'*

But it is on the western slope that the greatest number of ancient craters, lava-streams, and fields of scoriæ are met with. One of the principal volcanic foci is near the Salt Lake of the Mormons, where Mount Taylor rises to a height of 4,374 yards. This beautiful cone is surrounded by radiating lava-streams, which extend several miles.

'Several littoral or coast chains run parallel to the chain of the Rocky Mountains . . . . and in their northern portion are still the seat of volcanic activity.' Near the Gulf of California is the 'Volcanes de las Virgenes;' its last eruption took place in 1746. Rocks belonging to ancient volcanoes have been also found 'in the gold-yielding longitudinal valley of the Rio del Sacramento, and in a fallen-in trachytic crater called the Sacramento Butt. . . . . Further to the north, the Shasty or Tshashtl Mountains contain basaltic lavas.' . . . . The proper seat

* *Cosmos.*

of still-subsisting volcanic activity is found in the Cascade Mountains, where several peaks, covered with perpetual snow, rise to heights of 15,000 and 16,000 feet.*

The most remarkable are—Mount Saint Helen, a beautiful regular cone, always sending forth smoke from its crater—an eruption took place in 1842; Mount Reignier, which was active also at the same time; and Mounts Baker, Edgecombe, and Fairweather, still burning and covered with scoriæ.

The little isle of Lazare, near Sitka, in 57° N. lat., contains a volcano, the last eruption of which took place in 1796. In 1806 a lake was found in the crater, then in a state of inactivity. Warm springs rise in the neighbourhood.

In 60° N. lat. rises a giant volcano, Mount Elias. Its smoking summit can be seen fifty leagues from the coast. The chain of mountains, which up to this point turns from the north towards the west, falls back suddenly towards the south-west, forming the long peninsula of Alashka, continued by the long chain of the Aleutian Isles through the whole length of the North Pacific.

MacClure, in his voyage in the 'Investigator,' to discover the North-west Passage, points out, at the east of the mouth of the Mackenzie river (in 69° 57′ N. lat.) the volcanoes of Franklin Bay.

According to the description given by the missionary Miertsching, interpreter to the expedition, they were principally what are termed 'earth-fires,'

* *Cosmos*, vol. iii. p. 400 (Sabine's edition).

or emanations of Salses. Forty columns of smoke rose from conical elevations of clays of many colours. Great heat of the sea-bottom was perceived. At night luminous phenomena were seen from the ship, and at a great distance there was a strong smell of sulphur.

VI

VOLCANOES OF THE PACIFIC AND INDIAN OCEANS.

THE CIRCLE OF FIRE—VOLCANIC CHAIN OF THE ALEUTIAN ISLES AND KAMTSCHATKA—VOLCANOES OF JAPAN—VOLCANOES OF JAVA—DISAPPEARANCE OF THE PEPENDAJAN—BARREN ISLAND—ERUPTIONS OF TIMBORO AND GUNUNG-API—VOLCANIC ISLANDS OF OCEANIA—THE GALAPAGOS ISLES—THE MAUNA-ROA—VOLCANOES OF NEW ZEALAND—'EREBUS' AND 'TERROR'—VOLCANO OF BOURBON—ERUPTION OF THE DJEBEL-DUBBEH—THE RUINS OF SODOM.

THE CIRCLE OF FIRE.

If we imagine the north pole placed in the middle of Southern Europe, the terrestrial globe will then be divided by a corresponding equator into two hemispheres presenting remarkable contrasts.

The northern hemisphere will contain all the continents, the extreme part of Southern America excepted, and the continents will be grouped round Europe, better placed still as the centre of the globe. The southern hemisphere, on the contrary, will be nearly all seas.

Moreover, in the latter, covered by the immense Pacific Ocean, the volcanoes will be found in greater numbers than on the continental hemisphere. Before the recent period of its discovery, the real extent of the actual subterraneous activity of the globe was not known.

The long chain which adorns with its burning cones the western shore of America has already been described. Parallel to the opposite coast of Asia, and through the groups of islands which extend from Kamtschatka as far as New Zealand, appears a similar series, sometimes broader, and having lateral branches. Two volcanoes, near the south pole, appear to indicate that it continues across the continent, which covers probably a great part of the glacial zone; and that it is afterwards connected by the Shetland Isles, also covered with craters and lava, to the burning giants of the Andes.

Leopold de Buch has given the name of 'circle of fire' to this volcanic line, which forms one of the most characteristic features of the terrestrial surface. The Pacific Ocean is filled with numerous crater-islands, but those principally remarked are contained in two igneous zones—one extending from the Philippines to Easter Island, and the other from Japan as far as the colossal crater of the Sandwich Islands.

If, from a point distant from the earth, we could embrace the vast terrestrial hemisphere in a single glance, we should sometimes see in this volcanic region more than a hundred craters in eruption, grouped at night into one splendid constellation. The illustrious German geographer, Carl Ritter, adds some very interesting observations to the description of these fire-circles, which doubtless reveal immense fissures in the terrestrial crust.

'The force of upheaval,' says he, 'was exhibited formerly with a much more powerful activity in the entire basin of the southern seas. Indeed, independently of the islands that we see, others, still invisible, upheaved by thousands, are close under the surface of the water, under the form of shallows, rocks, and reefs; and, if the movement of the tides permits, they serve as supports to the superstructure of immense colonies of polyps and madrepores. But at the present time the expansive force of the subterranean vapour seems, in distributing itself over all these thousands of points, to become ineffectual to bring to the surface this submarine continent, the extent of which is not yet fixed by a series of sufficiently complete soundings.

'This action, applying itself to vast extents, and not only to isolated spots, again shows itself in the upheavals of the old and new world, which heap up their highest plateaux and their proudest mountains round the volcanic ring; whilst from the other side, towards the interior of the continent, the great plains descend into the North Atlantic Ocean and the vast Arctic depressions. The continental formation contrasts thus with the great insular formation, and both serve as a basis for past and future history.

'To the west of the mighty range of volcanoes of Oceania, and in their immediate proximity, extends the vast and beautiful country of Australia, which, deprived of any known volcano, could not be raised higher for want of the necessary force. Even the Great Barrier, so rich in corals and dangerous reefs, which rises between this continent and the long island of New Guinea, has not been able to emerge from the sea, or else has been replunged into the waves.

'This vast depression of a whole continent is continued also towards the north, between the Gulf of Carpentaria and the south-west of Malacca, along the isthmus of Sunda, pierced with such numerous straits. Beyond, the low lands of India, and Tonkin in Eastern China, extend as far as the central Asiatic plateau, which elevates, opposite the volcanoes of Japan, the not-to-be-crossed walls of the escarped coasts of Leaostong and of the Corea.

'An analogous phenomenon is seen in the two Americas. There also all the great depressions begin immediately beyond the Cordilleras, and the raised straight long plateau which this chain carries on its shoulders. Remarkable analogy! Here, as in the Australian continent, no volcano rises in these immense plains, the slope of which, like that of the rivers, descends from the exterior side of the volcanic circle. Scattered with a few groups of modest mountains, these plains sink and flatten from terrace to terrace as far as the Atlantic, while the interior side plunges perpendicularly into the Pacific.'

VOLCANIC CHAIN OF THE ALEUTIAN ISLES AND KAMTSCHATKA.

Between America and Northern Asia the bed of the sea forms a projection the volcanic energy of which is continually active. Nearly every point in the Aleutian archipelago has presented the phenomena of the appearance and disappearance of islands, of which the group of the Azores has given an example, a description of which we will give further on. More than thirty-four volcanoes are counted as having been in recent eruption. Mariners frequently feel on board their vessels the shocks of earthquakes which agitate these islands, and perceive smoke above most of their peaks.

At nearly a right angle with the Aleutian chain extends that of the volcanoes of Kamtschatka, to the number of fourteen, most of them being active. The principal of them is Klintchewskaja-Sopka, the height of which is 5,468 yards. Very abundant eruptions of lava have occurred in all of them, and there are also many places which have the greatest analogy with some parts of the Mexican plateau.

To the south of Kamtchatka, on the prolongation of its volcanic line, the Kourile Isles present ten actually in flames. We shall only mention them, as they are now rarely visited. In general the burning focus of the northern regions, of which we have spoken, and which is of such great importance to science, has not yet been sufficiently explored by travellers.

VOLCANOES OF JAPAN.

Near the isle of Jezo the naturalists attached to the expedition of 'La Pérouse' found a bay strewn with red porous lava and scoriæ. On the island itself rises seventeen conical mountains, which are for the most part extinct volcanoes. One of

Fig. 38.—Fousi-Yama, in the Gulf of Yeddo.

them is called by the Japanese the Mortar-mountain, on account of the deep depression of the crater, where there are some signs indicative of recent eruption. On the little island of Risiri, the volcanic peak of Langle rises from the sea to a height of 1,859 yards.

In the other great islands of Japan, seven active volcanoes are noticed—two in Niphon and five in

Kiousiou. The volcano Wunzen, which is as high as Vesuvius, lost its summit in 1793 during frightful explosions. The Fousi-Yama attains a height of 4,155 yards. It has a cone of remarkable regularity, truncated only near the summit, and having a vast oval crater. The upheaval of this mountain, venerated by the Japanese, who make frequent pilgrimages to it, in the year 286 before our era, is recounted by their historians. 'A vast extent of land sank,' writes one of them, 'in the country of Omi, a lake was formed, and the volcano Fousi appeared.'

To the north of Yeddo is Asama-Yama, the most central of the active volcanoes, at which occurred a very disastrous eruption in 1783. Among the small islands, two have the name of Iwo-Sima or sulphur isles: these are constantly smoking. Extinguished craters and cones of trachyte are frequent in all the mountain chains of Japan.

VOLCANOES OF JAVA.—DISAPPEARANCE OF PEPENDAJAN.

The Chinese island of Formosa, very rich in oil, contains four volcanoes, one of which, called Tschy-Kang, or Red Mountain, has experienced grand eruptions, and now possesses a crater-lake, filled with boiling waters. It marks the starting-point from which the lines of upheaval take a direction from north to south and even beyond the equator.

Here opens the most active part of the circle of fire. In the group of Southern Asia there are not less than 120 volcanoes, half of which have been lately in action; and it is probable that when the interiors of the great islands are explored, more will be found.

More than half of the forty-five craters of Java still vomit flames. The sea surrounding them is noted for its storms and tempests. There is sometimes in the air such a quantity of clouds charged with electricity that more than twenty waterspouts are seen at once. In this part of the torrid zone, the terrestrial fires rival the heat of the sun's rays.

Maha-Meru, the Sanscrit name of the largest of the volcanoes of Java, recalls the time when the Malays received Indian civilisation. It is a souvenir of Merou, the mythical mountain which in the Vedas represents the throne of Brahma.

The Gunung-Tengger is remarkable for its large crater, of circular form, which is nearly four miles in diameter. On the floor of the crater, 2,000 feet below the top of the wall, rise four eruptive cones, of which one only, the Bromo, has recently ceased sending forth flames. From 1838 to 1842 a lake formed there, the waters of which are warm and acid.

The Gunung-Pependajan suffered in 1772 the most violent eruption which has ravaged the island in historical times. Very different accounts are given of this frightful event. According to some, between the 11th and 12th of August, after the formation of a large luminous cloud, the mountain disappeared entirely in the bowels of the earth, and an extent of land seventeen miles long and seven broad was engulphed with it. Others say that the top of the volcano was destroyed by successive explosions, vomiting out cinders and enormous fragments on the surrounding country, where forty villages were buried. Two other volcanoes, situated the one 186 and the other 347 miles from Pependajan in a straight line, were in eruption at the same time, but several inter-

Fig. 39.—The Gunung-Tengger at Java.

mediate cones of the chain remained inactive. This fact indicates the complex character of the communication which must exist between the fissures of eruption through which volcanic matters make their exit. The Gunung-Guntur, or Thunder-mountain, made formidable noises during many consecutive years. Five great torrents of lava, the last dating from 1800, have flowed from the summit and reached

Fig. 40.—BARREN-ISLAND.

the foot of the volcano at different times. In the last eruption of 1800 it sent out, besides lava, an enormous stream of white, acid, sulphurous mud, doubtless proceeding from a solfatara, and which devastated the surface of a formerly fertile valley. Further on we shall have to mention other muddy eruptions, more frequent in Java than in any other part of the globe.

BARREN ISLAND.—ERUPTIONS OF TIMBORO AND GUNUNG-API.

The islands of Sumatra, Celebes, and Borneo, larger than Java, have comparatively fewer active volcanoes. On the first seven have been noticed, on the second eleven, and one only on the last. Nearly a hundred other volcanoes, half being still in action, are found scattered among the multitude of surrounding islands. The group of Nicobar and Andaman, a northern extremity of the volcanic chain of Sumatra, contains, according to Poulett-Scrope, the most remarkable type of the insular volcano, consisting in an active cone surrounded by the ramparts of an ancient crater in which the sea enters by a breach. This is Barren Island, called also by sailors the Deserted Isle. Its present form probably proceeds from an explosion which elevated a cone of very large dimensions. That which now remains is about 1,300 yards in height: loud eruptions succeed each other at intervals of ten minutes.

In 1638, the colossal cone, called the Peak, in the

Isle of Timor, disappeared entirely, and was replaced by an abyss now containing a lake. Until then this volcano, in continual activity, served as a lighthouse to mariners.

The Isle of Sambava is celebrated by a terrible eruption of its volcano, the Timboro, which took place in 1815. The following details are given in an account by Sir Stamford Raffles :—

'The eruption began on April 5, and was most violent from the 11th to the 12th, and did not quite cease until July. There were first detonations, which were heard at Sumatra, a distance of nearly 931 miles, and were taken for discharges of artillery. Three distinct columns of flame rose to an immense height, and the whole surface of the mountain soon appeared covered with incandescent lava, which extended to enormous distances; stones, some as large as the head, fell in a circle of several miles in diameter, and the fragments dispersed in the air caused total darkness. It is added that a waterspout accompanied the beginning of the eruption, and tore up the roofs, trees, and destroyed men and horses. The shore near the town of Timboro sank to a depth of six yards. The explosion lasted thirty-four days, and the abundance of the ashes expelled was such that at Java, a distance of 310 miles, they caused complete darkness in midday, and covered the ground and roofs with a stratum several inches thick. At Sambava also, the region in the vicinity of the volcano was entirely devastated, and the houses destroyed with 12,000 inhabitants. Thirty-six persons only escaped the disaster. The trees and pasturages were buried deeply under pumice and ashes. At Bima, at forty miles from the volcano, the weight of the ashes, &c. that fell was such that the roofs were crushed in. The floating pumice in the sea formed an island three feet in thickness, that the vessels could scarcely pass through.'

At the northern point of the Isle of Sanguir, the volcano of Abo, which, in 1711, covered a great number of villages with cinders, broke out all at once in the month of March 1856, and caused great disaster from the lava, cinders, stones, and torrents of mud it emitted.

The volcanoes of Celebes are at the north-east of the island. Near them are boiling sulphurous springs, the observation of which has deprived science of a distinguished savant. 'It was,' says Humboldt, 'into one of its springs, situated on the road from Sonder to Lamorang, that the indefatigable traveller and observer of nature, my friend the Piedmontese Count Carlo Vidua, was scalded to death.'

We shall mention, further, the Gunung-Api, or Fire-mountain, in the island of Banda, one of the Moluccas. This volcano is scarcely ever quiet, and from 1586 to 1820 twelve periods of violent eruptions have been noticed, throwing out streams of lava, scoriæ, and flames. In the month of June 1820 it emitted incandescent stones, as large, says one account, 'as the houses of the inhabitants.' These stones rose to a height of 1,300 yards above the island.

VOLCANIC ISLANDS OF OCEANIA.—THE GALAPAGOS.

The vast archipelago which extends from one tropic to another, in the midst of the Pacific Ocean, comprises more than a thousand islands. Their forms, generally annular, all seem to indicate volcanic origin; but the researches of the eminent naturalist, Darwin, have proved that the *attols*, or madreporic islands, are not, as was first believed, constructions raised by the zoophytes on upheaved craters. These prodigious masses of corals, which represent the work of centuries, crown ordinary mountains undergoing a very slow sinking. There are not really in the whole region so many volcanoes as in the single island of Java. Navigators have pointed them out here and there in the spaces between the attols. In general, a movement of the submarine bed, the inverse of that which lowers these attols, upheaves by degrees volcanic islands. Their height is not great: some craters only rise from 300 to 328 feet above the level of the sea. Several examples are given of periodical eruptions at short intervals. One of the most remarkable volcanic groups is that of the Galapagos Islands, distant only about 250 miles from the coast of America. It has been described in the following manner by Darwin, after he had visited it in the sloop 'Beagle,' commanded by Captain Fitzroy, who has since rendered such great services to meteorology:

'In the morning (September 17) we landed on Chatham Island, which, like the others, rises with a tame and rounded outline, broken here and there by scattered hillocks, the remains of former craters. Nothing could be less inviting than the first appearance. A broken field of black basaltic lava, thrown into the most rugged waves and crossed by great fissures, is everywhere covered by stunted sunburnt brushwood, which shows little signs of life. The dry and parched surface, being heated by the noonday sun, gave to the air a close and sultry feeling, like that from a stove; we fancied even that the bushes smelt unpleasantly. Although I diligently tried to collect as many plants as possible, I succeeded in getting very few; and such wretched-looking little weeds would have better become an arctic than an equatorial flora. The brushwood appears, from a short distance, as leafless as our trees during winter; and it was some time before I discovered that not only almost every plant was now in full leaf, but that the greater number were in flower. . . .

'One night I slept on shore, on a part of the island where black truncated cones were extraordinarily numerous; from one small eminence I counted sixty of them, all surmounted by craters more or less perfect. The greater number consisted merely of a ring of red scoriæ or slags, cemented together, and their height above the plain of lava was not more than from fifty to a hundred feet; none had been very lately active. The entire surface of this part of the island seems to have been permeated, like a sieve, by the subterranean vapours; here and there the lava, whilst soft, has been blown into great bubbles, and in other parts the tops of caverns, similarly formed, have fallen in, leaving circular pits with steep sides. From the regular forms of the many craters they gave to the country an artificial appearance, which vividly reminded me of those parts of Staffordshire where the great iron-foundries are most numerous. The day was glowing hot, and the scrambling over the rough surface and through the intricate thickets was very fatiguing; but I was well repaid by the strange Cyclopean scene.'*

THE MAUNA-ROA.

The most powerful focus of activity of Oceania is at the northern extremity of the archipelago in the Sandwich Isles, which are entirely volcanic. Hawaii, scarcely larger than Corsica, serves as a basis to a volcano 4,700 yards high, which consequently exceeds by 1,640 feet the Peak of Teneriffe. It is the Mauna-Roa, still burning in the middle of several extinct cones.

The craters of the summit of this volcano, the largest of which is nearly 4,370 yards in diameter, generally present a firm surface, composed of cooled lava and scoriæ, in the midst of which several orifices emit smoke. There is no cone of cinders; lava only is emitted in the eruptions. In 1833 and 1843 they lasted many weeks, and emitted streams five to seven miles broad. That of 1855 began by a brilliant jet,

* *Journal of Researches into the Natural History and Geology of the Countries visited during the Voyage of H.M.S. 'Beagle' round the World,* p. 337. By Charles Darwin, F.R.S.

Fig. 41.—CRATER OF THE MAUNA-ROA, IN HAWAII.

divided into thousands of drops, which rose about 500 feet above the summit of the dome. Immediately after, lava was ejected through an opening some 2,000 feet lower down, on the side of the mountain. This stream was enormous; it rapidly spread in the valley which separates the Mauna-Roa from the neighbouring volcano, the Mauna-Kea, attaining a breadth of three miles, which was soon doubled. The progress of this fiery stream was not arrested until the expiration of ten months, after reaching a length of 69 miles, in which it had carried with it entire forests.

The Rev. M. Coan, who then visited the mountain, relates that he crossed several times the hardened surface of the lava, under which it flowed in a liquid state, like the water in a frozen river. 'The superficial crust,' said he, in his account,* 'cracked with noise, emitting mineral vapours in a thousand places. On the edge were broken trees, half burnt, and falling into ashes on the hardened lava. . . . We passed several crevasses, through which we could see the fiery river, which precipitates itself, in its petrified canals, with a rapidity of several miles an hour. This lava was incandescent, and was from 25 to 100 feet in thickness; the openings or crevasses in the surface measured from 1 to 40 fathoms broad. We threw into these crevasses large stones, which, when they had touched the surface of the torrent, disappeared suddenly in flames. We could also see sub-

* Geological Society's Journal, 1856.

terranean igneous cataracts rolling over precipices from 10 to 65 feet deep.'

The lava is of an extraordinary fluidity, and Professor Dana has collected some curious facts on this subject. When it passes through the middle of forests many branches of trees retain particles of it, in such a manner that one sees suspended, here and there, stalactites similar to the icicles formed by the frost. Moreover, these branches, enveloped by the matter in fusion, remain intact; the bark is sometimes scarcely carbonised. It is supposed that they were wet at the moment when the lava reached them, and that the vapour, suddenly disengaged round them, preserved them.

Another effect of this fluidity is the production, by the lava projected in the air, of thousands of very fine glasslike threads, which the wind disperses all over the island. The imagination of the inhabitants transforms them into the hair of the goddess Pellé, the tutelary deity of the country. According to the legend, Pellé inhabited at first the Isle Mani, where a volcano, now extinct, Halea-Kava, probably also spread capillary glass.

Professor Dana has described several masses of lava of singular form, which he met with on the slopes of the volcano. They resembled petrified fountains, pillars, or straight bottles. Some were at least a hundred feet in height—the opening seen at top resulting from the explosion of gas, which had sent out liquid jets successively solidified.

The most remarkable phenomenon that the Mauna-

Roa presents is the crater of Kilauea, a vast lake of lava, situated on its eastern slope at 1,300 yards above the level of the sea.

'The longer diameter of the basin of Kilauea is 16,000 feet, and its shorter diameter 7,460 feet. The steaming, boiling, heaving fluid of this lava-pool does not, under ordinary circumstances, occupy the whole cavity, but only a space a little less than 14,000 feet long and rather more than 5,000 feet broad. There is a descent by a kind of steps down the crater margin. This grand phenomenon leaves on the mind a wonderful impression of stillness and solemn repose. The approach of an eruption is not here announced by earthquakes or subterranean noises, but solely by a sudden rising and falling of the surface of the lava, sometimes from a depth of 300 or 400 feet, to the upper margin of the basin, and the reverse. If, disregarding the enormous difference of size, we should be disposed to compare the colossal basin of Kilauea with the small lateral craters (circumstantially described by Spallanzani), situated on the declivity of Stromboli at four-fifths of its height, and of which the larger are only about 200 and the smaller about 30 feet across, we should remember the important distinction constituted by the fact, that these fiery openings of Stromboli (which is itself unopened at the summit) throw up scoriæ to a great height, and even pour forth lavas. Although the great lava-lake of Kilauea, which may be termed the lower and secondary crater of the active volcano of Mauna-Roa, may sometimes threaten to overflow its margin, yet it never has actually so overflowed as to give birth to a proper stream of lava. Such streams are, however, formed by the descent of the lava through subterranean channels, and by the formation of new openings of eruptions at a distance of 16 to 20 miles from the fiery lake at points much lower down. After such eruptions, occasioned by the pressure of the enormous mass of lava in Kilauea, have taken place, the surface of the lava in the basin sinks to a lower level.' *

There is generally remarked, all round the interior wall, a black border, which shows the height to which the lava rises before its running through the lower fissures. The Rev. S. Stewart has proved that when a fresh ebullition has broken the crust of the incandescent mass, the liquid surface hardens so rapidly in the air, that it becomes possible to walk on it a few hours after it has solidified on the surface. In the crater, the lava appeared to him scoriaceous and cellular, whilst near the exterior orifices of the mountain it was more compact and vitreous: he compares the first to the frothy surface of a liquid in fermentation, and the second to the same liquid drawn clear. It is remarkable that, during the most violent eruptions from the summit, the lava of Kilauea, to a distance only of 15 miles, remains in habitual repose. No sympathetic effect manifests itself between the two craters. It is known that hydrostatic laws do not permit a column of fluid to remain in one arm of a siphon 1,300 yards higher

* *Cosmos*, vol. iv. p. 375.

than in the other, so that one must conclude the absence of a subterranean canal joining the volcanic channels.

VOLCANOES OF NEW ZEALAND.

At the southern extremity of Oceania the two great islands which form this country present volcanic phenomena similar to those of Iceland. A

Fig. 42.—Thermal Springs of New Zealand.

chain of mountains, covered with eternal snows, crosses them, and contains many active volcanoes. Basaltic rocks, streams of lava of great extent, vast spaces filled with thermal springs, frequently appear in the magnificent views which these islands present to the traveller.

The principal volcanoes are found in the northern

island, along a fissure going from one sea to the other, and perpendicularly to the longitudinal chain of mountains. At the point of intersection rises the most active amongst them—the Tangariro, 2,200 yards high. The largest, Mount Edgecombe, now extinct, is situated on the sea-shore.

A line of lakes follows the chain of volcanoes. Round the largest, Lake Taupo, and at the centre of the island, we see, in a radius of two miles, the ground covered with solfatara and thermal springs, which, like the geysers of Iceland, form depôts of silica. The isthmus of Auckland, on which is situated the capital of the island, presents numerous traces of ancient volcanic activity, by the side of the region which it has powerfully fertilised.

'EREBUS' AND 'TERROR.'

From 1838 to 1841 the expeditions sent by France, England, and the United States discovered, beyond the great northern icebanks, a series of coasts, which appear to belong to a rather extensive continent. The mighty forces by which it has been upheaved in the middle of the deep seas of these regions are still in activity; gigantic volcanoes raise their heads above these frozen lands, and immense jets of flames coming from their craters illuminate the dark night which envelopes them during half the year.

The two first igneous mountains seen received the names of the vessels of the English expedition, 'Erebus' and 'Terror,' commanded by the intrepid

Fig. 43.—MOUNT EREBUS.

Captain James Ross. We find in the journal of surgeon MacCornick a description of the grand aspect of the new lands: 'On January 11, 1841, at 71° south latitude and 171° east longitude, the Antarctic Continent was seen for the first time. A chain of mountains with innumerable summits, reunited in distinct groups and covered with perpetual snows, appeared above the sea, and shone with magnificence in the sun. A peak, similar to an immense crystal of quartz, rises 2,600 yards high; another to 3,100 yards; and another of 3,280 yards. By the side of the white strata of ice numerous streams of lava and basalt descend towards the coast, where they terminate in abrupt promontories.

'The 28th, in lat. 77° and long. 167°, Mount Erebus was discovered—a burning mountain, enveloped in ice and snow from summit to base. From it issued a column of smoke, which extended over a great number of other cones, with which this extraordinary country is filled. The height of this volcano above the sea is 3,900 yards, and Mount Terror, an extinct crater which lies near it, is 3,900 yards. At its base is a cape, whence a barrier of ice extends westward, and stops all progress towards the south. We followed this perpendicular rampart for 300 miles.'

VOLCANO OF BOURBON.

The volcanoes of the Indian ocean are found in the vicinity of the great island of Madagascar, which itself contains active volcanoes, very little known up to the present time. At the northern extremity of the Mozambique canal, in the largest of the Comoroo islands is a cone in eruption. The Mascarine group, to the east, presents the most remarkable phenomena. The Mauritius is surrounded with a belt

Fig. 44.—The Peak (Isle of Bourbon).

of basaltic rocks. From its central plain are numerous currents of lava, which have cut a road towards the sea through many breaches. Cones rise up in different places, and as well as the principal mountain, called the Peak, have exhibited craters in eruption in modern times. There exists in the island of Bourbon a volcano which seems to have created it, and whose eruptions, very abundant and nearly continuous, never ceases to increase it. Extinct craters attest a volcanic activity more or less remote throughout the whole island. The peak now occupies the south-east part, and it is a position which it will most probably retain.

Fig. 45.—Crater of the Volcano of Bourbon.

Indeed, supposing that a primitive submarine eruption had formed the nucleus of the island at a very distant period, the cinders, sparks, and all the lightest materials, would have been driven by the wind to the north-east. The lava itself, feeling more or less the action of the waves, would also spread itself more easily with the wind, in which direction it would meet with fewer obstacles. All that could be detached from the lavas, and carried by the tides, formed on the same side a commencement of alluvium, and in time, the same causes continuing to produce the same effects, the primitive crater is nearer the sea on the side of the prevailing wind than on the opposite side. Nevertheless, at

Fig. 46.—The Grand Brûlé.

each eruption the cooled lava has formed round the crater whence it has issued an outer layer which, always rising and receiving new strata, has not been long in forming a mountain. When this mountain had attained a certain height, the crater was of great depth and dimensions, which the lava had to traverse and to fill before finding an issue. It is easy to understand that the expansion of the gas would then exercise on the lava, and on all the solid parts which enclosed it, a pressure as much stronger as the resistance was greater; and as the windward side had always been the least fed, it was on this side that the resistance should be overcome. New craters were thus opened, and will open successively, always in the vicinity of the sea,

Fig. 47.—The Crater of the Grand Brûlé.

and always on the windward side. These terrible convulsions, the opening of new craters and the formation of new mountains which are the consequences, explain perfectly the elevated peaks, the deep ravines, the immense circles which fill the whole of the interior of the islands, as well as the alluvium which extends to the sea on the one side, and the inaccessible ramparts which border it on the other. Thus is equally explained the superior quality of the soil, its greater depth, the more advanced decomposition of the lava in the part to leeward, and the contrasts pointed out by geology between this part of the island and the south-east side, which is evidently of more recent formation.

'The Grand Brulé, which now extends over several leagues, does not yet show any trace of vegetation. It is an inclined and unequal plain, filled with sharp irregularities, which rise and continually change form and aspect, by the effect of the streams of lava which annually furrow it, sometimes in one part and sometimes in another.

'This desolate country is destined to become in time a fertile land, and many years will not elapse before innumerable ferns, finding nourishment there, will prepare the light bed of humus, whence real forests should spring. The vicinity of the volcano of Bourbon proves the accomplishment of this metamorphosis with an inconceivable quickness.'*

ERUPTION OF THE DJEBEL-DUBBEH.—RUINS OF SODOM.

In its northern part, the Indian ocean washes coasts which were known in the most ancient times as foci of volcanic activity. We find in Arabian writings of the middle ages the mention of frequent eruptions, which took place to the south of Arabia, in the Zobayr chain of islands, in the straits of Bab-el-

* *Album de l'île Bourbon.* By Adolphe d'Hastrel.

Mandeb, in that of Ormuz, and in the eastern part of the Persian Gulf. The volcano situated near Medina vomited enormous torrents of lava in 1254 and 1276, but it has remained extinct since that period. The promontory of Aden is entirely volcanic, and the town itself is built at the bottom of a breached crater. In the Red Sea, the island of Djebel-Taïr is a volcano constantly in action. We will give a few details relative to a recent eruption of the Djebel-Dubbeh, which rises on the Arabian coast of this sea.

'In the night from the 7th to the 8th of May, 1861,' writes Captain Playfair, 'the inhabitants of Edd were awakened by a shock of an earthquake, followed by others, which continued with slight interruption during nearly an hour; at sunrise, a great quantity of white ashes fell on the town like rain; at nine o'clock these ashes changed their aspect, and seemed to resemble red earth. A short time after, this rain was so thick that the gloom became intense, and lamps had to be lighted in the houses. It became darker than the darkest night, and all the place was covered with ashes to the height of the knee. On the 9th, the rain of ashes diminished a little, and in the night columns of fire and thick smoke were seen to rise from the Djebel-Dubbeh, situated a day's march in the interior. Djebel-Dubbeh contained inhabitants, but not one had yet arrived at Edd on my departure from that place. They had never heard of a volcanic eruption; judge of the general consternation.'

According to another account, the noise of the volcano was heard as far as Perim, and at a great distance out to sea; vessels were enveloped in a cloud of ashes which recalled the thickest of London fogs. When the volcano could be approached, it was seen that a frightful catastrophe had taken place. The villages of Moobda and Rambo were buried under the materials thrown out by the crater.

The existence of vast basaltic formations and several craters has been demonstrated in the peninsula of Sinai.

Palestine also contains many vestiges of volcanic action; and Syria, in general, is subject to very violent earthquakes. The long valley watered by the Jordan, and occupied partly by the Lake of Tiberius and the Dead Sea, or Lake Asphaltite, very probably lies in a deep fissure of the crust. On the banks of these sheets of water, plenty of pumice, bitumen, and sulphur are found. The destruction of the towns of Sodom, Gomorrha, Seboim, Segor and Adama, buried, according to tradition, under a rain of fire, may be attributed to volcanic action. Recent researches confirm this conjecture.

'. . . . We entered,' says M. E. Delessert, 'into the desert and aridity, but this aridity was amply explained in this locality by the ground on which we trod; we were in the land of Sodom, and we were going to the extremity of the Dead Sea; to our right, we had a volcanic crater, washed by a tide, but easily recognisable by its perpendicular front and amphitheatre shape; it was Oued-ez-Zouera; to the left, the sea was narrower, and edged to the east by immense rocks nearly black; and lastly, before us an isolated mountain of about three miles in circuit, and near which we were obliged to pass, to reach the south, and afterwards to cross the immense plain which separates west from east. It is a mountain celebrated in history; we have no longer before us an ordinary land, fertile or not; it was a kind of salt crust, mixed with a little earth, on which the feet of our horses left a deep imprint; the mountain was in general of a yellowish tint and rounded form. At about one mile from its base, we made a *détour* to avoid a place where, a year before, a laden camel disappeared into a gulf about eighty feet deep, which suddenly opened; it was a light stratum which, softened by the continual rains, had again laid open the abyss which it covered. These details may give you an idea of the interest which attached itself to our march on such a treacherous soil; but the ground did not sink, and at eleven o'clock we passed the north angle of this mountain of salt. There is also a rather large ruin, composed of a mass of shapeless stones; it is Redjoum-el-Mezorhel; to the right, and covering a space of about one mile, are other ruins, but ruins like those of Engaddi, level with the surface, like those of a house would be that had been completely razed, and of which nothing remained but the foundations, showing certain angles which indicate the presence of old buildings; we saw a great number of these angles, important, since they prove the presence of a town, and that town Sodom.'

The name of this wicked town is preserved in that of S'doum, which the Arabs have given to the place. They call Djebel-S'doum the immense neighbouring salt-mountain, in which the winter rains have worn numerous fissures. A rock in the form of an *aiguille* detached from the mass, reminds one of the statue of Lot's wife.

'At sunset,' says a travelling companion of M. Delessert,* 'we had crossed New Sodom, and, passing between the two hillocks which cover the ruins of Segor, we entered the Oued-ez-Zouera, by which we were to regain the land of Canaan, and get to Hebron. Never shall we forget the magnificent spectacle which was presented to us when we had climbed the first outlier of the Canaan chain. A violent storm which came from the west had passed over these mountains, and, passing over the Red Sea, it was about to break over the plain of Moab. At sunset, the sky was perfectly free from vapour; to the east it was of a most gloomy tint; at the feet of the mountains of Moab, the sea seemed a vast sheet of molten lead; and the mountains themselves, black at the base, were red as fire mid-way to the top. All together we gave a cry of admiration; it was the conflagration of the five towns which recommenced again before our eyes.'

* M. de Saulcy. *Excursions on the Borders of the Dead Sea.* 1851.

VII

PREHISTORIC AND LUNAR VOLCANOES.

ANCIENT VOLCANOES OF FRANCE—BASALTS—PRIMITIVE ERUPTIONS—PHENOMENA OF CONTACT—INFLUENCE OF VOLCANOES ON THE ATMOSPHERE—THE UPAS VALLEY—LUNAR VOLCANOES.

ANCIENT VOLCANOES OF FRANCE.

The surface of our planet shows the most numerous and the most incontestable proofs of the primitive revolution which put it in communication with the still liquid materials of the interior, upheaved through deep crevasses, and solidified by the contact of the air. These materials, sent out by the prodigious force of vapours and gas, produced by the enormous subterranean heat, were also vomited forth by the craters of volcanoes, which, according to a just expression of Humboldt's, are, so to speak, the intermittent springs of the globe. The illustrious naturalist has remarked on this subject, how near the rich imagination of Plato was to these ideas, when this great philosopher assigned to the volcanic eruption and to the heat of the thermal springs a single cause, spread universally in the bowels of the earth, and symbolised by a river of subterranean fire, the *Pyriphlegethon.*

The ancient volcanoes, or prehistoric volcanoes, are met with everywhere. Hungary, Auvergne, Italy, Spain, Greece, England, present a very great number of extinct craters, whence radiate streams of lava, trains of scoriæ, and other products of volcanic activity. There are again found in different parts of the world, the same traces of ancient domination of the subterranean forces which still raise or tear the crust of our planet, in the midst of terrible convulsions and devastating eruptions. These phenomena, in appearance limited, nearly always extend, on the contrary, over vast regions, and thus give us the idea of the prodigious power which nature puts into play to accomplish its work.

We need not trace here, according to the recent discoveries of science, the history of old volcanoes. We shall confine ourselves to reproduce the principal features, making known the most curious particularities of the regions explored by geologists.

In France, the products of ancient volcanoes are seen on the coasts of the Mediterranean, in the Velay and Vivarais; but it is especially in Auvergne that the volcanic phenomena have left the most striking, and often the most singular marks of the disturbance of the surface at the different epochs, where the fiery matters are spread over the slopes and in the valleys. The mountains of Auvergne, which are connected with the Cevennes, contain three principal groups, Puy-de-Dôme and the mounts Dore and Cantal, on which more than a hundred craters are counted. The observations of geologists prove that in a late eruption from these craters issued torrents of lava, and materials more or less fluid, the cooling of which has produced volcanic monuments which so much attract the attention of travellers and naturalists.

BASALTS.

Among these monuments, the most remarkable are the basalts, rocks of igneous formation, that is to say, rocks emitted from the bosom of the earth in a fluid state, which spread themselves in sheets on the sides of the mountains, or in the valleys. This rock, very hard, black or of a bluish grey, is also sometimes greenish or red. The division of the basaltic masses into constant and regular forms—prisms, cubes, and spheres—constitutes one of the wonders of our volcanic region.

'The prisms of great length are often formed of pieces, which fit one in another, the lower side of each piece having a convexity which joins itself in a corresponding concavity of the upper extremity of the next piece. It has been remarked that in a bundle of prisms, thus articulated, the articulation is on the same line, that is to say, on the same level; also when, by a denudation, it is possible to see the plane of a basaltic surface thus divided, it resembles a large mosaic, which, in different localities, has been designated by the name of "*pavement*," or "*Giants' Causeway*." The northern coast of Ireland is particularly noted for the beauty and dimensions of the basaltic prisms that are there met with, and by the famous Giants' Causeway, which can be seen near Cape Fairhead. Fingal's Cave, in the isle of Staffa, to the west of Scotland, is not less celebrated on account of its majestic dimensions. The walls of this grotto, in which the sea is engulfed to the depth of 164 feet, are formed of regular vertical prisms, the

Fig. 48.—The Giants' Causeway.

height of them being 65 feet, which support a ceiling itself divided into prisms, laid in different directions.' *

In the Vivarais, on the borders of the little river of the Volant, a very beautiful basaltic causeway is found, which may be followed until it meets a stream of lava which, near Antraigues, flows from an ancient crater, where beautiful chestnuts flourish

* *Dictionnaire universel d'Histoire naturelle.* By C. D'Orbigny. Article 'Basalte,' by M. Constant Prévost.

in the midst of volcanic *débris*. The fertility of these disturbed countries, and their rich vegetation, present a picturesque contrast with the severe aspect of the arid regions devastated by fire. One of the most lovely basaltic colonnades of France is at Espaly, near Puy, on the banks of the river Borne. In some localities the columns of prismatic forms attain an elevation of 65 feet, and a diameter of one foot. The name of organ-pipes has been given to this group, and is one often employed to designate similar colonnades. The opposite side of the river has a very singular assemblage of basaltic prisms disposed in rays round a common centre, and forming an immense circle.

The basaltic depôts, rising from the bosom of the earth by straight chimneys or by long fissures, have everywhere the same mineralogical character. They are spread over the surface of the globe, principally in Iceland, Scotland, Bohemia, Germany, Italy, America, the Antilles, Teneriffe, the Isle of Bourbon, Saint Helena, Ascension, and in nearly all the islands of the South Sea. At Saint Helena, a basaltic needle of singular form is one of the most remarkable geological facts of the island. Its elevation, which is more than 65 feet, and its straight form, have obtained for it the name of the Chimney, but the horizontal position of the prisms gives it more the appearance of a pile of firewood.

The basaltic masses present great variety in structure. The different modes of division have been determined by the chemical composition of the mass, by the form of the stream, by the slow cooling, and, above all, by the application of a general law that we can here only mention, in virtue of which the work of the forces put into play by nature to produce any result whatever, is always the smallest possible. It must be remarked that the horizontal or vertical

Fig. 49.—Basaltic Mountains.

position of the prisms proceeds from the rock, which, in consolidating, arranges itself perpendicularly to the surfaces, by which the cooling is effected. All observations prove that basalts in sheets have been spread on nearly horizontal ground. The inclination of a great many of these sheets proceeds from the upheavals of the soil which followed their formation.

The colonnades of basalt furnish quarries where stones are found already shaped, and this particularity has exercised a certain influence on the primitive process of architecture. Nature, by a wonderful arrangement of these great rocky masses, united useful instruction to all those benefits which it lavished on the primitive peoples, and which it still offers, with great liberality, to savants who study it, and to artists who contemplate and love it. It is by seeing everywhere signs of order, that we have learnt to know, desire, and profit from the silent lessons which the flowers of the fields or the lava-rocks give us, on which is imprinted this natural geometry, the traces of which we again find in the midst of the immense convulsions which have cut out our valleys and upheaved our mountain chains.

PRIMITIVE ERUPTIONS.—PHENOMENA OF CONTACT.

Volcanoes are not only destructive agents, they produce new combinations of the substances contained in the circle of their activity, and, as we have seen, give them new forms. If, guided by science, we go back to the first ages of our planet, and if we try to represent the aspect of its surface in the numerous regions where volcanic chains rise, the influence of these chains on the present constitution of a great part of the earth's crust is easily comprehended. Terrible, nearly incessant eruptions not only covered the ground with cinders, scoriæ, fragments of rock, *débris* torn from the sides of the mountains, but they also modified the rocks already existing, either in the interior of the volcanoes, or at the passage of the lava-streams, transformed the lands deposited by the waters, and contributed to the formation of the mineral riches, which are heaped up in the crevices produced by the action of the subterranean forces.

The metallic veins are chiefly found near eruptive rocks, principally granite or other plutonic rocks, which, unlike the volcanic rocks, have risen in a pasty state in immense crevices, compared by Humboldt to valleys and gorges of very great extent. Thus, Mr. Darwin has observed in the Cordilleras of Chili granite in contact with sedimentary beds,* which are crossed by numerous veins of iron, copper, silver, and gold. Let us remark, in passing, that granite, one of the most widely-spread rocks, is always characterised by prismatic divisions, and presents also sometimes concentric divisions similar to those of certain basalts.* We may recall the fact that *sublimation* is a chemical operation by which the volatile parts of a body raised to a high temperature are collected. Metallic veins have probably been filled by this process, mineral substances subjected to an intense heat having been carried in a sublimated form into the fissures of the strata above them. Buckland relates on this subject the results of an experiment by the aid of which lead ore was produced by sublimation, in an earthen tube, the lower part of which was brought to a high temperature.

* The sedimentary earths have been deposited by the waters in horizontal beds, afterwards inclined, upheaved by rocks of eruption and earthquakes.

Metallic veins may also have been slowly formed by a sort of infiltration, the cause being electro-chemical action continued during a great lapse of time. Whatever it may be, it is certain that immense mineral riches are deposited in the strata which are found in contact with eruptive rocks. And to take still one example in the regions which are near us, we will mention the iron mines of the eastern part of the Pyrenees, which are met with in the calcareous beds in contact with granite.

Blast furnaces, in which forces act similarly to those which determine the chemical combination in the bosom of the earth, present us in their scoriæ with minerals formed artificially, identical with the most important simple minerals of which eruptive rocks are composed. An eminent chemist, M. Mitscherlich, has proposed the artificial reproduction of minerals by analogous means, and has obtained, among other mineralogical substances, garnets and rubies. Chemists have since succeeded in reproducing the corundum, sapphire, and a certain number of other beautiful crystals similar to those which are developed on the surfaces in contact with igneous rocks and sedimentary beds. It is thus that we may recognise the action of the plutonic forces in the diamond districts of Brazil and the European slopes of the Oural. The emerald, ruby, sapphire, topaz, garnet, and opal are also met with near old lava-streams, and sometimes in the beds of rivulets which traverse volcanic countries. These fine stones are not found in distant countries only; the bed of a rivulet which flows near the Puy in Velay, at the foot of the old volcano of Croustet, produces, in France, a good quantity of rubies and sapphires.

In some regions, where the ground is strewn with brilliant grains detached from the crystallised rocks, a curious fact has been observed—the ants' nests are filled with these grains; we quote on this subject the following passage from an account of a well-known explorer, M. Jules Marcou: 'There exists on the high plateaux of the Rocky Mountains a species of ant which, instead of using wood and remains of vegetables to build its house, employs only small stones of the size of a grain of maize. Its instinct teaches it to choose the most brilliant pieces. Also the nest is often filled with magnificent garnets and grains of very clear quartz.'

INFLUENCE OF VOLCANOES ON THE ATMOSPHERE.

Observations relative to volcanoes, and researches on the interior heat of the globe, have helped to explain the presence in the northern glacial regions of the vegetables of the torrid zone, which are found in a fossil state. At the time when hot vapours were emitted everywhere from the interior springs, the temperature of the different zones, much higher in the lower and more uniform atmospheric strata, brought forth and developed the luxuriant vegetation, the remains of which, buried at the bottom of lakes, gulfs, and ancient seas, have been converted into pit-coal. The thickness of the coal seams, which sometimes attains sixteen yards, indicates the exuberance of this primitive vegetation, favoured both by a very high degree of temperature and the gaseous exhalations which render the air improper to the respiration of animals, but which furnished plants with a nourishment much more abundant. From this single fact we may judge that vegetation analogous to our crawling moss then attained a height of 200 feet. The volcanoes and thermal springs were still an abundant source of carbonic acid. It is certain that this gas, from which vegetation borrows the necessary carbon which we again find in the coal, must have been, in those distant periods, contained in our atmosphere in very great quantity, and has been submitted, by a series of upheavals and volcanic eruptions, to the influences which the interior of the globe still exercises on its composition. The presence in many coal districts of rich beds of ferruginous ore is remarkable.

It is to this circumstance, so favourable to the establishment of iron foundries, that England owes in great part its industrial power, resulting from the construction and prodigious work of the innumerable machines which 'row, pump, dig, carry, draw, raise, hammer, spin, weave, and print.'

THE UPAS VALLEY.

Carbonic acid, one of the most widely-spread and abundant bodies in nature, being heavier than air, often accumulates in low places and in the inferior parts of a great number of cavities in volcanic countries, such as the *Grotto del Cane* of Naples, where, as we have seen, the animals which breathe it are asphyxied in a few minutes. In Java, a crater called the *Guevo Upas*, or the Valley of Poison, 650 yards in circumference, also possesses a celebrity founded on the reports which attributed to the innocent emanations of the upas tree, the juice of which is used to poison arrows, the effects produced by the carbonic acid.

The following description explains the dull aspect of this strange place. 'The use of the upas was formerly general in all the islands, but the introduction of fire-arms has now banished it to a few savage tribes, who take refuge in the volcanic mountains with which the island is filled. These volcanoes are igneous or muddy, their unforeseen eruptions often cover large spaces with lava and mud. Sulphurous acid, and silicious springs, petrifying all the objects near, spout up from the ground. Sometimes from the top of a hill the astonished traveller all at once discovers a valley without vegetation, calcined by the sun. Skeletons of animals of all kinds lie over the ground; their posture proves that they have been seized suddenly when full of life; the tiger the moment when it had seized its prey; the vulture when he alighted on these carcases to devour them. Thousands of insects, ants, coleoptera, cover the soil: it is a valley of death. Carbonic acid escapes by the fissures in the ground, and, in virtue of its specific weight, it remains invisible at the bottom of the valley: an analogous phenomenon to that of the *Grotto del Cane* and of *Dunsthoehle,* near Pyrmont. Man alone may cross these valleys of death, on account of his head rising above the bed of gas. The Indians, who pass through the defiles of the Himalaya, some being 5,500 yards above the sea, attribute to the emanations of the surrounding plants the uneasiness and difficulty of breathing really due to the rarity of the air; in the same way the Javanese are said to ascribe to the emanations of venomous trees the disastrous effects of an irrespirable gas.' *

LUNAR VOLCANOES.

The surface of our satellite is scattered over with a great number of mountains, having nearly all the form of a circular wall, in the middle of which is a depression. Laplace recognised evident traces of eruptions. He added that the formation of new spots and luminous points, many times observed in the dark portion, even indicate volcanoes still in activity. It is to them that he attributed the aerolites which fall from time to time on our globe.

New researches have considerably modified these ideas. They attribute the sight of the spots of light to optical illusions. The surface of the lunar world, drawn with the greatest care by astronomers, and even for some time past photographed, does not appear in any way changed, and a theory of aerolites, different from that of Laplace, now prevails. But if

* *La Plante et sa vie.* By Schleiden.

recent eruptions cannot be proved on the globe that accompanies us in our journey round the sun, we find numerous proofs of the existence of an epoch when the reaction of the interior of this body on its superficial crust was extremely violent. When we compare the relief of the surface on the earth and on the moon, it is surprising to find the disproportion between the mountains. They are relatively much higher on our satellite, where we count twenty-two which exceed the altitude of Mont Blanc (5,249 yards); the mountain called Dœrfel, 8,314 yards high, is only 218 yards less than the highest peak in the Himalayas. This extension of elevation seems in keeping with the diminution of gravity, which is calculated to be six times less on the moon than on the earth.

To see the cavities well, a time of observation between the first or the last quarter must be chosen. The circular ramparts then cast their shadows opposite to the sun, at one period inwards, the other time outwards. We are soon struck with the idea of a perfect analogy between these lunar formations and our terrestrial volcanic formations. The outside of the protuberance rejoins the plain by a moderate slope, whilst the interior perpendicular rock is extremely abrupt. In the central part of the bottom of the cavity are often seen eminences which represent very well our volcanic cones whence proceed lava and ashes. In many localities, according to Herschel, decisive marks of stratification,* proceed-

* Disposition of the collected matters in parallel beds.

Fig. 50.—IDEAL LANDSCAPE IN THE MOON.

ing from the successive deposit of the ejected materials, have been distinguished. We shall not be astonished at the proximity of the crateriform protuberances, if we state that more than two thousand are visible on the visible surface.

In fig. 50 we have endeavoured to give an idea of the scenery presented near one of these mountains. The great number of sharp points which rise up on the buttresses and the surrounding country give a strangeness to the spectacle. They are very high, and remind us of our basaltic columns. There is one near Mount Ligustinus which is ten times higher than Strasbourg Cathedral. When we saw it for the first time it was enveloped in shadow, and its point only was lighted up by the sun.

We should be mistaken if the analogy which suggests itself at the first sight conducted us to assimilate all the lunar craters to our volcanoes. It is a very small number only, on the contrary, which can be compared to them, when we take their dimensions into account. An observer placed on our satellite and furnished with an excellent telescope would scarcely see the terrestrial craters. The crater of the Caldiera del Fogo in the Island of Palmas, the largest, according to Humboldt, is only five miles in diameter, whilst on the moon we see a multitude of circumvallations much more extensive. The diameter of Mount Clavius, for example, is 140 miles, those of eight others are comprised between 69 and 113 miles, then come twelve circles of 55 miles on an average. We must, then, imagine vast areas as large as two or three English counties; in fact, like Bohemia. This last country, surrounded by mountains, would represent them very well, only the ramparts of the lunar-walled craters are real cliffs, scarcely crevassed and rising at a bound to a height which the top of Mont Blanc only reaches after a great development of declivities and outlyers.

It is only the black points seen on the sides of some protuberances which may be considered as craters of extinct volcanoes. The circles themselves have probably been produced by a process differing, at least in its gigantic proportions, from the volcanic eruptions with which we are acquainted. Let us imagine, in the first selenological periods, powerful elastic gases freeing themselves by a series of internal chemical reactions of the lunar mass, and finding themselves stopped by a stratum of very resistant miry substances still viscous enough to possess the faculty of extending. Numerous bubbles, acquiring large dimensions by reason of this elasticity, would project from the surface, and during some time the interior pressure of the gases dilated by the heat would sustain the vault, perhaps as large as Clavius. But as soon as the cooling commenced it would want support, it would break up and scatter its *débris* at the bottom of an immense abyss. What would then remain? Precisely the appearances that we observe now: all the supported parts sunk as far as the abrupt borders; mountains of rocks split at the centre, on an even disc the level of which is below that of the surrounding region. This last circumstance is characteristic of all the circles. The excavation of Mount Newton is so deep that the bottom of it is never lighted up either by the earth or the sun.

Among the fluids emitted from the fractured domes were probably many vapours which have condensed, and even been transformed into solid substances; for the observations of astronomers do not yet prove the existence of a lunar atmosphere. They attribute to these vapours the deposit of the bright pellicles which seem to clothe some borders of the circles. According to Maedler, they are due to gaseous streams, which have vitrified a portion of the surface, producing there luminous bands, which are seen disposed in rays round many of the mountains. In certain points, like Tycho and Copernicus, there are more than a hundred, and they uninterruptedly cross the circumvallations and the surrounding black spots, thus furnishing an element to establish the chronology of the upheaval of the surface. We must suppose that large conflagrations occurred above the centres, towards which the currents converged, as in all cases. None of these bands stand in relief, as they are only visible about the full moon, and do not throw any shadow during the other phases.

An English astronomer, Mr. Hooke, obtained an artificial imitation of the lunar cavities by heating calcareous mud until the steam, in the form of great bubbles, forced its way through the surface. In our terrestrial volcanoes, the upper stratum of matter in fusion sometimes rises by the elasticity of the subterraneous gases as far as the edges of the crater, but the dome sinks as soon as the gases have made a passage. It is known that there exist in America great extents of land which are hollow underneath, and which are in fact real bubbles. If we wish to compare the lunar surface with that of

Fig. 51.—Extinct Volcanoes of Auvergne.

our globe, we must in imagination suppress the sedimentary earths and the seas which cover the latter. Many circles, now filled up, would then appear. In Auvergne there are some very large, which are still entirely sunken, although the granite which forms them is mixed up and disappears in a great number of points under thick beds of vegetable earth. The one seen in the island of Ceylon is 43 miles in diameter. In Oceania several madreporic islands appear to be supported on similar circles. 'We can then figure to ourselves,' as remarked by Humboldt, 'our satellite nearly like what our earth was in its primitive state, before it was covered with sedimentary beds rich in shells, gravel, and diluvium, due to the action of the tides and streams. Scarcely can we admit that there exists in the moon beds of conglomerates,* and of detritus formed by friction.'

* Aggregation of different mineral substances.

VIII

EARTHQUAKES.

EARTHQUAKE OF LISBON—EARTHQUAKE IN CALABRIA—EARTHQUAKE AT RIOBAMBA—VOSTITZA—EARTHQUAKE IN THE INTERIOR OF A MINE—SUBTERRANEAN NOISES—GUADALOUPE—PADANG—DESTRUCTION OF MENDOZA—GEOGRAPHICAL DISTRIBUTION OF EARTHQUAKES.

EARTHQUAKE AT LISBON.

THE following letter, addressed to one of the members of the Royal Society of London, by Mr. Wolfall, surgeon, is extracted from the 'Philosophical Transactions':—

'Lisbon, Nov. 18, 1755.

'. . . . On the 1st inst., about forty minutes past nine in the morning, was felt a most violent shock of an earthquake. It seemed to last about the tenth part of a minute, and then came down every church and convent in town, together with the King's Palace, the magnificent opera-house joining to it; in short, there was not a large building in town that escaped. Of the dwelling-houses, there might be a quarter of them that crumbled, which, at a very moderate computation, occasioned the loss of 30,000 lives. The shocking sight of the dead bodies, together with the shrieks and cries of those who were half buried in the ruins, are only known to those who were eye-witnesses. It far exceeds all description, for the fear and consternation was so great, that the most resolute person durst not stay a moment to remove a few stones off the friend he loved most, though many might have been saved by so doing; but nothing was thought of but self-preservation. Getting into open places and in the middle of streets was the most probable security. Such as were in the upper stories of the houses were in general more fortunate than those who attempted to escape by the doors, for they were buried under the ruins, with the greatest part of the foot-passengers. Such as were in equipages escaped best, though their cattle-drivers suffered severely. But those lost in houses and the streets are very unequal in number to those that were buried in the ruins of churches; for it was a day of great devotion, and the time of celebrating mass. All the churches in the city were vastly crowded. . . . .

'In about two hours after the shock fires broke out in three different parts of the city, occasioned by the goods and the kitchen fires being all jumbled together. About this time also, the wind, from being perfectly calm, sprung up a fresh gale, which made the fire rage with such fury that, at the end of three days, the city was reduced to cinders. Indeed, every element seemed to conspire to our destruction, for, soon after the shock, which was near high water, the tide rose thirty feet higher in an instant, and as suddenly subsided. Had it not so done, the whole city must have been laid under water. As soon as one had time for recollection, nothing but death was present to our imaginations.'

After stating the apprehension of a pestilence and the fears of a famine, the writer proceeds: 'The third great dread was that the low villainous part of the people would take an advantage of the confusion and murder and plunder those few who had saved anything. This in some degree happened; upon which the King gave orders for gallows immediately to be placed all round the city; and after about 100 executions, amongst which were some English sailors, the evil stopped. We are still in a state of the greatest uncertainty and confusion, for we have had in all twenty-two different shocks since the first, but none so violent as to bring any houses down in the outskirts of the town that escaped the first shock; but nobody yet ventures to lie in houses; and though we are in general exposed to the open sky, for want of materials to make tents, and though rain has fallen several nights past, yet I don't find but the most delicate people suffer these difficulties with as little inconvenience as the most robust and healthy.' . . . .

Mr. Wolfall, in his second letter, states that the time the earthquake lasted was between five and seven minutes. 'The first shock was extremely short, but then it was as quick as lightning succeeded by two others, which, in the general way of speaking, are mentioned altogether as one shock.'

The oscillation of the terrestrial surface which

produced this terrible disaster was not a local event. It was felt over a vast area, said to be greater than four times the size of Europe. It was in Portugal, Spain, and in the northern part of Africa that the first shock was of the greatest violence. The port of Setubal, a few leagues from Lisbon, was also covered by an enormous wave; and at Cadiz, some high walls close to the shore were carried away by the sea, which rose 65 feet above its ordinary level. In Morocco, many towns were devastated, and thousands of inhabitants perished. On the western border of the Atlantic, in the little Antilles, where the tide scarcely exceeds 29½ inches, the water all at once became entirely black, and rose to a height of more than 22 feet. At the same moment, the Swiss lakes, those of Sweden, and the sea which washes the Norwegian coast, were violently agitated, whilst perfect calm reigned in the atmosphere. Many streams of water were diverted. The interception of the thermal springs of water in Tœplitz was also noticed, but they appeared again a short time afterwards of the colour of blood. They inundated the town, and struck the inhabitants with awe. To explain this singular phenomenon, we must admit a subterranean shock, changing the direction of the waters, and causing them to pass through a bed of red ochre. It is related also that Vesuvius, then in full eruption, suddenly became quiet at the moment of the commotion.

A field of action so extended shows that the forces which produced this immense earthquake were manifested deep in the interior of our planet, and not on the surface only. They probably give rise to waves of circular shocks, transmitted one after another, through elastic mineral masses, to the superficial stratum. The propagation of this movement takes place in a manner analogous to that of sound. Gay-Lussac very well describes this effect in the following passage: 'The earth, so many centuries old, still keeps an internal force, which agitates the entire mass. Most of the mountains rising from its bosom must leave vast cavities, which have remained empty, at least they have only been filled by water and gaseous fluids. It is erroneously that Deluc and many other geologists make use of these voids, which they imagine are prolonged in long galleries, in order to account for the propagation of earthquakes to a distance. These phenomena, so grand and so terrible, are very strong sonorous waves, excited in the solid mass of the earth by some commotion which is propagated with the same quickness as sound is propagated. The movement of a carriage on the pavement shakes the largest buildings, and is communicated through considerable masses, as in the deep quarries under Paris.' *

When in the terrestrial undulations the extreme limit of the elasticity of the bodies is exceeded, and ruptures take place, the crevasses give passage to gases, which in their turn produce movements of translation. This effect was observed in many places in Portugal, and particularly in the rocks of Alvidras,

* *Annales de Chimie et de Physique*, t. xxii.

which remained covered during several days with a thick cloud of smoke.

What is the cause of the internal movements of the globe which give rise to such grand and terrible phenomena? An eminent English geologist, Mr. Mallet, sees it in submarine eruptions, at the end of which water penetrates by open canals to the surface of the lava. There follow, then, according to him, violent explosions, the rebound of which, transmitted in every direction, constitutes an earthquake. Mr. Poulett-Scrope brings forward another hypothesis. Mineral masses situated deeply would suddenly increase in temperature on receiving an addition of heat from the interior focus, and their dilatation would produce successive rents in the adjacent rocks, and at the same time 'undulatory pulsations.'

The learned professor, M. Daubrée, makes water, derived both from the atmosphere and the ocean, play a part in these grand mechanical effects; he admits that it penetrates not only in the warm subterranean regions by extensive fissures, but also by slow infiltration, resulting from the porosity and capillarity of the rocks. Laboratory experiments have shown him that like infiltrations are produced even in presence of very strong interior counter-pressure. 'Do not these experiments,' he asks, in a note,* 'touch the fundamental points of the mechanism of volcanoes and other phenomena, which have been attributed generally to the development

* *Compte rendu de l'Académie des Sciences*, January 28, 1861.

of vapours in the interior of the globe, principally earthquakes, the formation of certain thermal springs, the filling of metallic veins, as well as various cases of metamorphism of rocks? Without excluding the original water, and some kind of initial constitution supposed to be generally incorporated in the interior and molten masses, do not these experiments show that infiltrations, descending from the surface, may also intervene in such a manner that many of the deep portions of the globe would be daily receiving and dispensing by a most simple process, differing, however, from the mechanism of the siphon and ordinary springs? A slow, continuous, and regular phenomenon would thus become the cause of the rude and violent manifestations comparable to explosions and ruptures of equilibrium,'

EARTHQUAKE IN CALABRIA.

Calabria is the part of the great Mediterranean volcanic region most exposed to earthquakes. Those which occurred at the end of last century were especially disastrous; we shall here bring together a few details of the effects produced by the shocks of February 5 and March 28, 1783.

In an extent of about sixty square leagues all the surface of the country was devastated. Of 375 towns or villages, 320 were completely destroyed, and the others preserved very few of the houses intact. The ravages were extended as far as Sicily. At Messina, an enormous conflagration was added to the fall of

Fig. 52.- EARTHQUAKE AT MESSINA (1785).

the edifices, and the terror of the inhabitants was inexpressible. Along the Straits, the bottom of the sea sank several yards; the coast was inclined, and was torn by numerous cracks. The promontory, which formed the entrance to the port, disappeared in an instant. In the middle of Calabria, in the pretty town of Polistena, rich and well peopled, the greater part of the inhabitants perished under the ruins. Not a wall remained standing.

The French geologist, Dolomieu, who was travelling in Italy, thus describes the scene of this catastrophe: '. . . . I saw Messina and Reggio; I could not find there a house that was habitable, and which did not require to be taken down to the foundations; nevertheless the skeleton of these two towns still remains, and it may be seen what they have been. Messina still retains, at a certain distance, a perfect image of its ancient splendour. Each recognised his house or the ground on which it stood. I had seen Tropea and Nicotera, in which are very few houses which have not been much damaged, and many of which have entirely given way. My imagination did not go beyond the misery of these towns. But when, placed on a height, I saw the ruins of Polistena, the first town of the plain which presented itself to me; when I contemplated the heaps of stones which had no longer any form, and which could not give me the least idea of what the town had been; when I saw that nothing had escaped destruction, and that all was brought on a level with the ground, I felt a sentiment of terror, pity, fright, which for a few moments suspended all my faculties . . . .'

The ground opened in all parts, often in long crevasses, some of which were 500 feet broad, and more than 1000 yards long. Some were isolated, others bifurcated. Sometimes they were crossed by a series of perpendicular fissures. There were some which radiated round a centre, like the cracks in a broken pane of glass. Most of them remained open after the commotion; others, opened at the moment of the shock, were afterwards closed, crushing between their walls houses, trees, and men, which they swallowed up.

In many parts of Calabria, and especially round the town of Rosario, a curious phenomenon was seen. Circular cavities, similar to small cisterns, were opened here and there in the ground, and were filled with water and sand. On digging round them they were found to have the form of a shaft, which ended in a narrow and deep canal.

Enormous portions of land were transported to great distances, from the tops of the mountains into the plains. The town of Terranova, built above narrow valleys, was divided into portions, which fell after having been turned upside down. Fourteen hundred inhabitants were buried in the ruins. Houses were precipitated perpendicularly in a gulf 150 yards deep. The materials brought together by the earthquake stopped the course of a river, and formed a largish lake. In other places, brooks thus intercepted cut themselves new passages, and became

Fig. 53.—CALABRIAN PEASANTS ENGULPHED BY CREVASSES (1785).

torrents, the ravages of which were as disastrous as the shocks themselves. The centre of the shocks appears to have been under the territory of Oppido, where it made immense crevasses. The town gave way entirely, and a vast gulf opened on the slope of a neighbouring hill, swallowing up farms, warehouses, men, and herds.

The mountain of Aspromonte was subjected to great disturbance. The villages built on the steep rocks were detached with them and fell into the valley. In some places the land slipped, keeping intact the trees and plantations.

We will add the frightful disaster at the rock of Scylla. After the first shock, the houses and gardens were crushed by the detached rocks of the neighbouring heights. The Prince of Scylla persuaded a great part of his subjects to take refuge in the ships to get away from the earthquake. Towards midnight another shock was felt. The sea, rising suddenly 19 feet, was precipitated many times on the shore, drawing with it the vessels, which either went to the bottom or were broken against the rocks. Many were found at a great distance in the interior of the land.

EARTHQUAKE AT RIOBAMBA.

During this catastrophe, which took place at the beginning of the month of February 1794, the disastrous disturbances observed in Calabria were reproduced in the Andes on a much grander scale. Humboldt, who visited America at this time, collected on the spot many facts from the lips of the survivors; we will give the most important.

At Quito was seen a prodigious number of shooting stars a short time before the first shock. The same phenomenon preceded the earthquake of Cumana in 1766. Humboldt adds on this subject, that on a certain night the volcano of Cayambe appeared, for an entire hour, as surrounded by meteorites.* The inhabitants of Quito, frightened at this apparition, walked in procession, to appease the divine wrath. The disturbed space round the town of Riobamba, which was entirely devastated, comprised the whole high volcanic plain of Quito, between Tunguragua and Cotopaxi. A muddy eruption of the Moya destroyed forty thousand Indians on the same plateau; but that which establishes the influence of the commotion to a much greater distance is the following fact, observed near the town of Pasta, sixty miles distant. The volcano which dominates it had vomited for three months a high column of smoke; at the moment of the first shock it disappeared entirely. Subterranean communications then began, and were suddenly established between the two igneous foci.

'The earthquake was neither announced nor accompanied by any subterranean noise. An immense detonation, now designated by these single words, *el gran ruido*, was produced only eighteen or twenty minutes later, under the two towns of Quito and Ibarra, and was not heard at Tacunga or Hambato, or on the scene even of the disaster.'

The following phenomena were also noticeable:—

* This of course was only a coincidence.—Tr.

Fig. 54.—EARTHQUAKE AT SCYLLA (1785).

'Fissures, which alternately opened and closed, so that persons partially engulfed were saved by extending their arms, that they might not be swallowed up; portions of long trains of muleteers and laden mules (*recuas*) disappearing in suddenly opening cross fissures, whilst other portions, by a hasty retreat, escaped the danger; vertical oscillations, by the non-simultaneous rising and sinking of adjoining portions of ground, so that persons standing in the choir of a church, sixteen feet above the pavement of the street, found themselves lowered to the level of the pavement without being thrown down; the sinking down of massive houses, with such an absence of disruption or dislocation that the inhabitants could open the doors in the interior, pass uninjured from room to room, light candles, and debate with each other their chances of escape, during two days which elapsed before they were dug out; lastly, the entire disappearance of great masses of stones and building materials. The old town had possessed churches, convents, and houses of several stories, but in the places where they stood, we found, on tracing out among the ruins the former plan of the city, only stone heaps of from eight to twelve feet high.' *

VOSTITZA.

The trembling of the ground, even in the absence of subterranean noises, produces, when felt for the first time, a very peculiar effect, which has been well depicted by Humboldt. 'I do not think that this

* *Cosmos*, p. 271, vol. iii. pt. i.

impression is produced by the recollection at the moment of the dreadful images of destruction which historic relations of past catastrophes have presented to our imagination; it is rather occasioned by the circumstance that our innate confidence in the immobility of the ground beneath us is at once shaken; from our earliest childhood we are accustomed to contrast the mobility of water with the immobility of the earth; all the evidences of our senses have confirmed this belief; and when suddenly the ground itself shakes beneath us, a natural force of which we have had no previous experience presents itself as a strange and mysterious agency. A single instant annihilates the illusion of our whole previous life; we feel the imagined repose of nature vanish, and that we are ourselves transported into the realms of unknown destructive forces.'

We have experienced these impressions during a violent earthquake of which we were witness in the month of May 1844. Our brig was at anchor before Vostitza, in the Gulf of Corinth, and we had landed to visit a Greek convent in the neighbourhood. The monks had received us in a room from which there was a magnificent landscape. While we were resting ourselves, strong shocks all at once shook the walls, and we saw enormous rocks rolling down the slopes of the mountains, whilst the air was filled with thick dust. The astonished monks were at prayer in the church, and caused the bells to be rung. We made vain efforts to persuade them to leave the building, already showing large crevices. During our return to the shore, we heard the subterranean thunder. Strong shocks had been felt on board the brig. The next day, we learnt that a village, a few leagues distant, had half its houses thrown down.

Vostitza was subjected to a fearful earthquake on April 23, 1817. The lower part of the shore and Cape Aliki were plunged beneath the waters, already become very warm. In a few minutes the town was reduced to a mass of ruins. The same shock was violently felt at Patras and as far as the Elide.

Towards the commencement of the thirteenth century before our era, an earthquake, which struck Greece with fear, swallowed in the bosom of the waters the town of Helice, situated near the sea, not far from Vostitza (the ancient Ægium). In the time of Ovid, who mentions this catastrophe, its edifices could still be seen under the waves. According to the legend, the Acheans who inhabited Helice broke faith with some suppliants who had taken refuge in the Temple of Neptune and killed them, and the anger of the gods immediately showed itself by the catastrophe which annihilated the town. 'Ordinarily,' says Pausanias on this subject, 'earthquakes which from time to time disturb certain countries, are announced by prognostics which precede them, such as continual rains or long drought, or disarrangement of the seasons, or the darkness of the sun, or the dryness of the fountains, or whirlwinds which root up the largest trees, or celestial fires which traverse the vast extent of the air, leaving behind them a long train of light, or new stars which appear suddenly and fill us with fear, or pestilential vapours which come from the bosom of the earth. Such are the signs by which Heaven warns men.

Near Helice, the town of Bura, where rose, in front of Delphi, the temple of Hercules, in which priests attended to whom was attributed the gift of foretelling the future, was entirely destroyed by the same commotion. The ravine in which its ruins were found is remarkable on account of the strange and picturesque forms of the great rocky peaks separated by deep crevasses.

EARTHQUAKE IN THE INTERIOR OF A MINE.

An engineer, M. de la Torre, who was in one of the copper mines in Cuba during a terrible shock of an earthquake which converted the town of Santiago into a heap of ruins, in the month of November 1852, thus relates his impressions: 'I was in the gallery number 132 of St. John's pit, superintending the work of a squad of twenty-four men. We were preparing the augers when we heard a noise as extraordinary as it was fearful, which made us fear the gallery would give way; we felt at the moment the ground rise and fall at the same time, throwing us at different times from one wall of the gallery to the opposite one. We looked for death as inevitable; but it seemed prudent to sit down to escape instant destruction. The lanterns fell from the walls from which they were suspended, and all was darkness. The supporting wood cracked, making a noise like that of a furnace fed with green fuel; the infiltration of the water increased in a wonderful way; it appeared to us, although we were in a mine, as if we were under a tree the bushy foliage of which, filled with dew, had been vigorously shaken by the storm, or rather by the hand of God. At the same time we smelt sulphur, and heard the noise of stones scattered from the upper parts of the mine into the lower ones. As I have before said, we were in the thickest gloom; there only remained one distant light, which served the more to show us the horror of our situation. We were all together and we dare not speak; in fact, we found ourselves literally between life and death. The noise lasted more than four minutes, although the shocks had already ceased. We decided, with some hesitation, to go, and when we had our feet placed on the ladders, another shock was felt; it would have certainly overthrown us if we had not been looking for it. After a thousand terrors we had the happiness to reach the opening of the mine. The joy which we experienced cannot be expressed. . . . .'

SUBTERRANEAN NOISES.

It has been remarked that the dull sounds which often accompany earthquakes do not increase in the same degree as the violence of the movements of oscillation. At the time of the great shock which destroyed Riobamba, a complete silence reigned. At other times the subterranean thunders are heard without there being any commotion.

'The nature of the noise also differs greatly: sometimes it is rolling, and occasionally like the clanking of chains; in the city of Quito it has sometimes been abrupt, like thunder close at hand, and sometimes clear and ringing, as if obsidian or other vitrified masses clashed, or were shattered in subterranean cavities. As solid bodies are excellent conductors of sound, which is propagated, for example, in burnt clay with a velocity ten or twelve times greater than in air, the subterranean noise may be heard at great distances from the place where it has originated. In the Caraccas in the grassy plains of Calabozo, and on the banks of the Rio-Apure, which falls into the Orinoco, there was heard, over a district of 2,300 square (German) miles, a loud noise resembling thunder, unaccompanied by any shaking of the ground; whilst, at a distance of 632 miles to the north-east, the crater of the volcano of St. Vincent, one of the small West Indian Islands, was pouring forth a prodigious stream of lava. In point of distance, this was as if an eruption of Vesuvius should be heard in the north of France. In 1744, at the great eruption of Cotopaxi, subterranean noises, as of cannon, were heard at Honda near the Magdalena river. Not only is the crater of Cotopaxi about 18,100 English feet higher than the Honda, but these two points are separated from each other by a distance of 436 miles, and by the colossal mountain masses of Quito, Pasto, and Popayan, as well as by countless valleys and ravines. The sound was clearly not propagated through the air but through the earth, and at a great depth. During the violent earthquake in New Granada, in February 1835, subterranean thunder was heard at Popayan, Bogota, Santa Martha, and Caraccas (when it lasted seven hours without any movement of the ground), and also in Hayti, in Jamaica, and near the lake of Nicaragua.

'These phenomena of sound, even when unaccompanied by sensible shocks, produce a peculiarly deep impression, even on those who have long dwelt on ground subject to frequent trembling. One awaits with anxiety that which is to follow the subterranean thunder. The most striking instance of uninterrupted subterranean noise unaccompanied by any trace of earthquake is the phenomenon which is known in the Mexican territory by the name of "the subterranean roaring and thundering of Guanaxato." This rich and celebrated mountain city is situated at a distance from any active volcano. The noise began on January 9, 1784, at midnight, and lasted above a month. I have been enabled to give a circumstantial description of the phenomenon from the report of many witnesses, and from documents of the municipality which I was permitted to make use of. From January 13 to 16 it was as if there were heavy storm-clouds under the feet of the inhabitants, in which slow-rolling thunder alternated with short thunder-claps. This noise ceased gradually, as it commenced; it was confined to a small space, for it was not heard in a basaltic district at the distance of only a few miles. Almost all the inhabitants were terrified, and quitted the city, in which large masses of silver were stored; but the most courageous, when they had become somewhat accustomed to the subterranean thunder, returned and fought with the band of robbers who had taken possession of the treasures. Neither at the surface, nor in mines, 1,600 English feet in depth, could the slightest trembling of the ground be perceived. In no part of the whole mountainous country of Mexico had anything similar ever been known before, nor has this awful phenomenon been since repeated. Thus, as chasms in the interior of the earth close and open, the propagation of the waves of sound is either arrested in its progress, or continued until it reaches the ear.' *

GUADALOUPE.—PADANG.—MENDOZA.

We will complete our series of accounts by a few extracts relative to three violent earthquakes of recent date.

On February 8, 1843, at thirty-five minutes past ten in the morning, with a calm atmosphere, a subterranean growling and very strong shocks, suddenly cast fear among the populations of Martinique and Guadeloupe. The first of these islands, disturbed by a similar plague in the year 1839, had little to suffer this time, but Guadeloupe was the most frightful spectacle of ruin and desolation. Its capital, Pointe-à-Pitre fell down entirely in a few seconds, and a conflagration breaking out in the

* *Cosmos*, vol. i. p. 195–6.

Fig. 55.—INCIDENT DURING THE EARTHQUAKE AT SUMATRA (1861).

midst of the rubbish finished the work of destruction. Immense crevasses, whence came whirlwinds of vapour and flames, swallowed up hundreds of victims. More than two thousand deaths were recorded. The principal industry of the country was destroyed in a moment: there only remained standing three sugar mills out of sixty that were established round Point-à-Pitre. Nearly all the standing harvest was lost.

A Frenchman established at the port of Padang has given the following description of the earthquake which took place in 1861, in the southern part of the volcanic isle of Sumatra: 'The earthquake began by a commotion which was felt on February 16, at seven in the evening, and which lasted about 115 seconds. Thanks to the peculiar construction of our houses the evil was limited, although the extreme violence of the motion of the ground made us think that some of them would not hold out. All the inhabitants fled with cries. I felt as if I were on the bridge of a vessel beaten by a tempest, and experienced all the symptoms of sea-sickness.

'The establishment of Singkel, on the extreme frontier of the Dutch possessions near the kingdom of Acheen, disappeared under the water, in consequence of the sinking of the peninsula on which it was built. The sea now covers the place on which were the fort and government magazines. The garrison was saved.

'At Polo Nyas, the sea, dashed on the shore by a violent submarine commotion, completely razed the fort as well as the establishment of Lagondie, and carried away forty-nine soldiers and Malay inhabitants when it retreated. The shocks were so strong that the most robust men were violently thrown on the ground.

'On the side of Gunung-Sitalie entire villages were soon nothing more than a heap of ruins, and a great number of the inhabitants were buried under the *débris*.

'On the western coast of the same island the ground sank in many places, and rose in others; coral islands rose from the bosom of the waters, others, on the other hand, disappeared. Hundreds of inhabitants died in the midst of these sudden convulsions.

'At Baros and Siboga the earth opened, springs of boiling water spouting out in different places. Eye-witnesses relate that here and there the ground opened and closed alternately, as if the earth were bent under the effort of volcanic work which was going on in its depths.

'The coast of Acheen was ravaged by the sudden invasion of the sea, which, penetrating into the interior of the land, overturned houses, trees, and harvests, and in receding carried away a great number of inhabitants.

'In the Batoa isles the sea, upheaved by an irresistible force to a great height, penetrated into the interior of the country, annihilating all that it met with in its passage; then, retiring with the same rapidity, it carried off 700 inhabitants from a single island, leaving nothing behind it but a frightfully ravined surface, where the eye looked in vain for a vestige of the rich vegetation which covered it a few hours before.

'The earth has not, so to speak, ceased to tremble since the evening of February 16; we have each day observed a greater or less number of shocks. The Merapi, the crater of which has not given signs of life for five years, emitted at this time thick columns of smoke; and from the Talang and Singaland came dull detonations.'

A few days after the Sunda archipelago had been thus overturned, one of the most terrible catastrophes of which history has made mention destroyed the town of Mendoza, situated in a beautiful position at the foot of the eastern side of the Cordilleras, on the route which leads from Buenos Ayres to Valparaiso. One minute in the night of March 20 sufficed to entirely overturn and to transform it into a vast heap of ruins, the highest of which were not six feet above the ground. The evening before an immense blue and red meteor crossed the heavens, lighting up a vast extent of country, and travelling slowly from east to west. At some distance from Mendoza the volcano of Aconcagua was in eruption.

'Never, in the memory of man, has a town been surprised with such violence, and without the earthquake being preceded, at least a few seconds, with these distant and subterranean rumblings, which give time either to fly or to throw ourselves into the arms of those we love and bid them an everlasting adieu. Nearly always the animals forebode disaster, and announce it to man by their agitation. This day, in less than four seconds, more than 17,000 persons were buried beneath the ruins. Frightful noises succeeded, terrifying cries, fearful howlings of men and crushed animals; the light of the conflagration spread with rapidity, a thick dust was diffused in the atmosphere, and the sky was darkened as in the darkest nights.'*

We borrow from the same source a touching episode which happened to a Frenchman, M. Tesser, a rich hotel-keeper, established with his family at Mendoza. 'One of his intimate friends wandered among the ruins. His eyes were dry, he could shed tears no longer. He stopped on the site of the hotel. After having tried in vain to recall the old arrangements, he was retiring—his heart filled with sighs, thinking of the honest man and this family he had loved so well—when he perceived, through the shapeless mass of girders and calcined stones, M. Tesser's dog, which moaned: he approached it. The poor animal, the two hind legs and part of the body of which were crushed, forced itself, in spite of its sufferings and weakness, to scratch with its front paws, and uttered from time to time a plaintive howl. As soon as it saw this friend of its master's come near, it exerted itself and howled louder. The friend understood that Tesser must be under this rubbish, and hoped he was not dead. He ran to fetch some persons, and with their help, after much labour,

* M. Ernest Charton, in the *Magasin pittoresque*, vol. xxxiii.

he indeed discovered the body of poor Tesser; his arm and left leg, lying under the beams, were broken, his mouth and eyes full of earth, but he still breathed. Before trying to disengage his limbs they washed his face, which seemed to relieve him; and, without saying a word he instinctively put his right arm towards his dog, who drew himself to him, and died a few moments afterwards.

'Tesser scarcely was in a state to pronounce any words before he asked where his family was. Alas! all had perished in the great disaster. Hearing this answer he closed his eyes with despair; then, making a fresh effort, he pronounced the name of his little girl, and showed with his finger a separate place, where he had put her to bed. Some of the people, in compassion for his grief, although without hope, made further search; others occupied themselves in dressing his broken limbs. A few minutes afterwards, those who were rendering him this service saw him suddenly raise himself up—he gave a cry—they brought him his daughter still living. A beam had fallen across the bed of the child and had protected it; but she was seriously wounded in the head; she had also her eyes and mouth filled with dirt, and she was exhausted with hunger. They placed both of them under a tent against a tree, and they remained there more than two months, less near to life, it seemed, than death. Tesser embraced with his unbroken arm his daughter—his only good on earth—his only hope after so many calamities.'

The centre of the shock appears to have been

under the town; the villages which surrounded it, although damaged, suffered little relatively.

Valparaiso and its vicinity are frequently the scene of earthquakes; but for at least a hundred years there has not been any on the side of Mendoza, and it was generally expected they would not be felt on the other side of the Cordilleras. The largest part of the town of San Juan, also situated at the foot of the Andes, forty leagues to the north of Mendoza, was destroyed at the same time, and 3,000 persons perished. At 130 leagues to the east, the church of Cordova fell down. The shock was also felt at Buenos Ayres, but did not cause any havoc.

GEOGRAPHICAL DISTRIBUTION OF EARTHQUAKES.

It has been remarked that the undulations which succeed each other in earthquakes have generally a constant direction, and this probably following the line of propagation in the interior of the earth. Sometimes, however, shocks in a certain direction alternate with others in different ones. In the earthquakes of Caracas in 1811, and Chili in 1822, the shocks from north to south were crossed, from time to time by others with a perpendicular direction. It also happens that compound shocks result from different shocks which are produced simultaneously.

The velocity of propagation is variable, and depends on the nature of the traversed surface. In the case of the earthquake of Lisbon it has been established, from the facts stated, that the velocity was five times greater between the coasts of Portugal and those of Holstein than along the Rhine. From Lisbon to Glückstadt, separated by a distance of 295 miles, the shock travelled 2,777 yards a second—that is, 1,175 yards less than sound travels in a cast-iron tube.

'When such a wave proceeds in a regular course along a coast, or at the foot of, and parallel to, the direction of a mountain chain, interruptions at certain points have sometimes been remarked, and continue for centuries; the undulation passes onward in the depths below, but it is never felt at those points of the surface. The Peruvians say of these upper strata which are never shaken, that they form a bridge. As the mountain chains themselves appear to have been elevated over fissures, it may be that the walls of these cavities favour the propagation of the undulations moving in their own direction; sometimes, however, the waves intersect several chains almost at right angles, an example of which occurs in South America, where they cross both the littoral chain of Venezuela and the Sierra Parime. In Asia, shocks of earthquakes have been propagated from Lahore and the foot of the Himalaya (January 22, 1832) across the chain of the Hindoo Coosh, as far as Badakschan or the Upper Oxus, and even to Bokhara.

'The range of the undulations is sometimes permanently extended, and this may be a consequence of a single earthquake of unusual violence. Since the destruction of Cumana on December 14, 1797, and only since that epoch, every shock on the southern coast extends to the mica-slate rocks of the peninsula of Maniqnárez, situated opposite the chalk hills of the main-land. In the great alluvial valleys of the Mississippi, the Arkansas and the Ohio, the progressive advance from south to north of the almost uninterrupted undulations of the ground, between 1811 and 1813, was very striking. It would seem as if subterranean obstacles were gradually overcome, and that the way being once opened the undulatory movement is propagated through it on each occasion.'*

The examination of earthquakes leads us to divide the globe into different regions. We can distinguish those in which they are violent and frequent from those in which they are only very insignificant in their effects and rare. Maps constructed with such data would be of great utility when they are based on a long period of observation, and furnished with detailed information on the stability of different countries. We shall confine ourselves here to the most general indications.

The part of the globe most exposed to earthquakes includes the Mediterranean and the adjacent countries, Asia Minor, the Caucasus, the Caspian Sea, and the mountains of Persia. They are connected with the volcanic region of Central Asia, the principal focus of which seems to be near Lake Baïkal. The Asiatic continent is subjected, to a great extent, to earthquakes; but, with the exception of the shores of the

* *Cosmos*, vol. i. p. 199, Sabine's translation.

Red Sea and the Barbary coast, Africa is entirely exempt. In another direction, except the terrible shocks which devastate the countries situated to the west of the Andes, the chain of the Andes itself, the Antilles, and the borders of the Gulf of Mexico, the phenomenon is very rare on the American continent.

It has generally been noticed in Europe that more earthquakes take place in autumn and winter than in the spring or summer. In a memoir presented to the Academy of Sciences, M. Alexis Perrey, who has for a number of years collected all the documents relative to these phenomena, arrives at the conclusion that their frequency increases at the syzygies, and when the moon is in or near perigee. The shocks, according to him, are also more frequent when the moon is in the vicinity of the meridian than when it is 90° distant.

The facts that we possess relating to the earthquakes which have agitated France before the year 1000 are very uncertain. They include the celebrated earthquake of 468, which destroyed Vienna in Dauphiny, and for which St. Maurent, bishop of that town, instituted the *Rogations*; that which in 842 lasted for seven days in the north of France; and those of 801, 829, and 950, which were nearly general over Europe.

From 1000 to our time we find nineteen earthquakes which have overturned public edifices or houses. The west, the north-west, and the north of France have been less subjected to them than the south-east, which lies nearer the active volcanic focus of Italy. A seventh of the total number of earthquakes has been observed in Provence and in a zone which lies along the Pyrenees, although the surface of these two regions is scarcely equal to the twentieth part of France.

IX

MUD VOLCANOES—SPRINGS AND WELLS OF FIRE—THERMAL SPRINGS.

MUD VOLCANOES—SALSES OF TAMAN AND TURBACO—SPRINGS AND ISLANDS OF MUD—BURNING MOUNTAINS—SPRINGS AND WELLS OF FIRE—THE CAUCASUS—FIRES OF BAKOU—WELLS OF BITUMEN—THERMAL SPRINGS—GEYSERS OF NEW ZEALAND—THE TE-TA-RATA—INFLUENCE OF THERMAL SPRINGS.

MUD VOLCANOES.

A GREAT number of salses appear to be connected with extinct volcanoes, and to represent their last phase. The enormous masses of mud produced by their eruptions are in some places considerable enough to form mountains. Sometimes still active volcanoes give birth to this phenomenon, as was seen in 1797 near Quito. The eruption began by an earthquake which shook the country over an extent of 170 leagues from south to north, and of 140 from west to east. At the centre of this region, numbers of villages were overturned and buried under the mud thrown from the top of the volcanic mountains.

We have already given a description of this catastrophe, which did not leave one house standing in the vast space shaken round the town of Riobamba. Torrents of mud escaped from the base of the volcano of Tunguragua, and formed streams which in the valleys rose 600 feet in height. The mud deposited by them barred the course of the rivers and formed lakes. These muddy torrents frequently proceeded from the melting of the ice and snows, or again from the condensation of the enormous quantities of vapour, which frees itself from the mouth of the volcano and falls in rain, mixed with ashes drawn with it.

At the north-west extremity of the Caucasus chain the peninsula of Taman and the eastern part of the Crimea present a pretty good number of hills which are evidently nothing but ancient mud-volcanoes.

There was an eruption of one of the salses of Taman on February 27, 1793. After strong subterranean detonations, a column of fire, half veiled in a thick vapour, rose up many hundred feet, accompanied with an abundant emission of mud and gas. In Iceland numerous mud-springs dart out of the middle of the little basins similar to craters. On the American continent, one of the most remarkable groups of salses is situated near the charming village of Turbaco, two and a half miles from Carthagena in New Grenada. The description of it has been given by Humboldt, and more recently by M. Vauvert de Méan. The *volcancitos*, to the number of eighteen to twenty, rise on a desert plain, in the midst of a great forest of palm-trees, which plain is bounded on the horizon by the colossal snow-covered mountains of Santa-Marta. The eruptions of gas and mud flowed, as at Taman, from the tops of little truncated cones formed of clay earth; they are six and eight yards high, and sixty to eighty yards in diameter at the base. The upper part of the funnel is filled with a liquid mud, constantly agitated by the ebullition of large bubbles of gas, which ascend with violence. Near the openings are heard at intervals hollow noises which precede the eruptions. The observations made since Humboldt's time prove that the gaseous emanations undergo chemical changes, and the same fact has been established for the salses of Taman. In 1839 a great eruption of flames and a convulsion of the soil accompanied the disappearance of the volcanic cone on Cape Galera-Zamba, eight miles from Carthagena. The straight tongue of land which formed the cape was thus separated from the continent by a channel thirty feet deep. In the month of October 1848, a new and terrible igneous eruption occurred at the place of the rupture; an island was upheaved from the bottom of the neighbouring sea, and disappeared a short time afterwards. Everything tends to the belief that the submarine volcano of Galera-Zamba is the principal focus of the phenomenon of the salses in the province of Carthagena, where there exist hundreds of cones vomiting saltish clay over a surface of 400 square leagues. More than fifty *volcancitos* similar to those of Turbaco now surround (in some cases at a distance of sixteen to twenty miles) the submarine gas-volcano of the Galera-Zamba.

In Java, and many other islands of the Indian archipelago, there exist similar salses to those of Turbaco. The gigantic volcanoes of Java frequently vomit torrents of mud which devastate the country, and appear to proceed from subterranean depths, or sometimes from the mixture of great quantities of ashes with the contents of the crater-lakes, lined in the interior with a stratum impenetrable to water, formed by ashes and conglomerates. Some are found also in granite, basalt, and other hard and massive rock. The formation of these great cup-like basins is attributed by Mr. Poulett Scrope to the powerful explosions of enormous bubbles of gas which are formed at the surface of a reservoir of very liquid lava, when a mass of vapour, at a high degree of tension, rises from the depths, and causes the lava to be in a state of ebullition. Dana at Hawaii, and Darwin in the Galapagos, have observed bubbles several yards in diameter, and it is not impossible that these bubbles are sometimes reunited in one colossal blister on the surface of the lava.

During the eruption of the Guntur, at Java, in 1800, a torrent of white acid sulphurous mud, doubtless proceeding from a solfatara, devastated the whole surface of a large valley.

SPRINGS AND ISLANDS OF MUD.

In a remarkable essay * on the rôle played by the liquid elements in the interior of the terrestrial crust, a learned observer, Mr. R. Thomassy, has collected a number of facts which tend to demonstrate the presence of an enormous quantity of pluvial waters in the fissures, crevasses, and cavities produced by subterranean disturbances. These waters, due either to superficial overflowing or to evaporation, penetrate and circulate in the bowels of the globe, which scarcely ever opens under the effort of volcanoes without water, either in a liquid state or in the state of vapour, being emitted at the same time as the fire.

We have already quoted the opinion of M. Daubrée on the possibility of a capillary infiltration through porous materials. The *Études sur le métamorphisme*, by the same learned physicist, contains also the following passage: 'In the volcanic exhalations there is a body which at first does not attract attention, because, under the impress of old ideas, it seems quite inert, especially in presence of the minerals, the formation of which is in question. It does not exist in minimum quantities, like the vapour of which we have just spoken; it is, on the contrary, both the most abundant and constant product of eruptions in every region of the globe. . . . . We can only know the masses situated at a certain depth, by what the volcanoes bring to the surface. Now, these ejections all contain, without

* *Essay on Hydrology*, by R. Thomassy.

exception, water, either combined or mixed; we have, then, a right to think that water plays an important part in the principal phenomena which emanate from the depths.'

M. Thomassy insists upon the action of water in limestone countries, where it produces a multitude of very deep natural excavations. In Greece, in the plain of Mantin, which forms a basin without issue, one sees, after great rains, torrents engulfed and disappearing through the caverns and fissures of the mountains. The waters thus absorbed must flow out somewhere, and they sometimes reappear in springs. Springs of this kind, overcharged with sediment, and seeming to proceed from the masses of water swallowed up in the vast limestone formations of Missouri, are met with in great numbers at the mouths of the Mississippi, where they produce the remarkable phenomenon of mud-islands. M. Thomassy quotes the description of one of these islands given by Professor Forshey:—

'Its length is about 600 feet, and the maximum of its present height 7 feet 4 inches. Not far from the eastern point is a salt-spring, which constitutes the principal characteristic of this island, and explains its formation. When we approach it, a cone is perceived from 2 to 3 feet high, by 50 at the base, from the summit of which continually escapes a lead-coloured mud, to which are added, from time to time, emissions of gas. The mud flows slowly on the sides and adds to the deposits, which are always increasing. This increase continues until the elevation thus formed attains about 7 feet above the surrounding waters. The spring then stops, but the eruption continues in a less elevated place, when it begins the same kind of work. The surface of the island exhibits traces of many similar monticules.

'In a sounding of the spring made by the aid of an improvised wooden bridge, the lead reached a depth of 25 feet. Similar soundings reached elsewhere 82 feet, but the thickness of the mud held the lead, which was lost by the breaking of the line. As to the water, which has a very salt taste without any other mixture, it disposes of its salt by evaporation, and the entire island shines with crystals thus deposited on its clay.'

The formation of mud-islands is due, according to M. Thomassy, to the force of upheaval exercised by sheets of subterranean water, more or less elevated above the mouths of the river. The muds are remarkable for their adherence; and the soil which they deposit, nearly always firm and solid, gives a particular character to the islands, which indicates their true geological place. 'Sentinels of terra-firma, advanced out at sea, and grouped around the mouth of the river, they offer halting-places to the shifting alluvium, and intercept the floating wood, which would have been dispersed by every wind of the gulf. Now those, once grounded on the banks, very soon encourage deposits of every kind. The little islands thus increase at each fresh arrival, and as all act in like manner, the rapidity of the development which is observed in the delta of the Mississippi' is explained.

The original struggle of water and fire is one of the fundamental principles of geology, and this struggle continues still. But taking M. Thomassy's point of view, we see that these two elements, contrary in appearance, unite in the work of creation, that their influence is equal in the formation of the globe, and that, instead 'of insisting on their ancient antagonism, it is their union that we must try to bring out.'

FIERY MOUNTAINS.—SPRINGS AND WELLS OF FIRE.

Springs of inflammable gas, fiery mountains, were in olden times the origin of the legends which, according to the true thought of Humboldt, were our first presentiments of the truth. Each place in which similar phenomena are produced, the ancient poets have peopled with monsters—terrible giants, whose powerful movements cause both subterranean convulsions and the eruptions that follow them. Thus, for example, the volcano situated on the Cragus, a high promontory in Lycia, was guarded by the Chimæra, a fantastic monster, who, from his gaping mouth, vomited a whirlwind of flames. Strabo speaks of the high tops of the Cragus, and Pliny quotes it as one of those natural lighthouses over the heights of the Mediterranean, which served to guide the first mariners.

The Chimæra situated on the coasts of Caraminia, near Deliktash, the ancient Phaselis, is now a source of gas constantly burning.

Strabo quotes a similar phenomenon, observed in Cappodocia, near the celebrated town of Cæsarea, situated at the foot of the ancient volcano Argæus, the highest among all the mountains of Asia Minor: 'The neighbourhood of Mazaca (Cæsarea) is sterile, and not susceptible of cultivation, for at the surface the soil is sandy, and at a certain depth rocks are met with. At a short distance from the town, one enters a vast plain, many stadia in extent. It is devastated by fire, and furrowed by excavations vomiting flames, which oblige the inhabitants of the town to go some distance to buy provisions.'

THE CAUCASUS.—FIRES OF BAKOU.

'The study of popular myths cannot be separated from the geography of volcanoes and their history; often these two orders of facts reciprocally explain each other.' The poetical legend of the dragon of the Hesperides is again met with at the other extremity of the old world, where the dragon of Colchis also indicated the existence of a volcanic region. Pherecydes of Samos, quoted by Humboldt, mentions in his Theogony, 'that Typhon when pursued tried to save himself on the Caucasus, that the mountain was burning, and that he took refuge thence in Italy, where the isle Pithecusa (Ischia) was thrown, and, so to speak, melted around him.' This fable is evidently a remembrance of the volcanic eruptions of Caucasus, as well as the account of Apollonius of Rhodes, who places in this mountain the rock of

Typhon, on which this giant was struck by a thunderbolt by Jupiter, son of Kronos (Time).

These mythical personifications of phenomena resulting from the combination of the atmospheric and terrestrial elements, these allusions to the physical catastrophes of which the old world had been the scene, indicate not only, according to the just observation of M. Guigniaut,* that creation develops itself by struggle and combat, as well as by union; they show also the advent of a new general principle, which, through Jupiter, governs the world, evolved from the primitive chaos, 'both in space and time.'

The mud-volcanoes and fires of naphtha are arranged in certain lines, which indicate the extent and connection of these phenomena. In the high valley of Kinalughi, 7,800 feet above the sea, shine 'the perpetual fires of Schagdagh.' We will describe here only the naphtha-springs of Bakou, situated at the south-east extremity of the Caucasian chain.

'Naphtha is a kind of bitumen, a very inflammable liquid. The ground on which Bakou is built is full of it. If one introduces a stick rather deeply in the earth in any place, and brings a light to the orifice that has been made, there is immediately a jet of gas. Vegetation does not flourish around Bakou, not because the soil is not fertile—it is superabundantly heated by subterranean fires—but because water is absent, so that at Bakou a garden is a princely luxury.

'From all time, the town of Bakou has been con-

* *Theogony of Hesiod.*

sidered as a sacred town by the Guebres. There is a convent of Parsees situated near Bakou, which contains the famous sanctuary Atesh-Gah, where the eternal fire burns. The priests are only three in number, and came from Delhi; they have another convent at Bombay. Persecuted by the Mahometans since the year 655, the Parsees are proscribed and

Fig. 56.—Fires of Bakou.

dispersed. They never eat anything that has lived or spill blood. These poor people are the most gentle and most inoffensive of men. Those who are at Atesh-Gah live peaceably under the protection of the government of Russia.

'We arrive at a large plain. The fires come from openings irregularly placed. In the midst rises an embattled edifice; from each battlement proceeds a tuft of flames. A more intense focus, composed of five fires, crowns the highest cupola. In the interior, the spectacle is imposing; everywhere fire comes out of the earth; under the central cupola the altar is covered with flames. . . . .

'It only remains for us to see the fires of the sea. On the morrow we went in a boat, on a beautiful night, as far as the naphtha springs, which we instantly knew by their odour. One of the sailors, furnished with tow, lighted some, and threw it into the sea, at a place where it appeared boiling. In an instant, all the surface of the sea was lighted up to an extent of 130 feet. We went further on, repeating the same experiment, and the conflagration extended. We sailed on an ocean of fire. What a display! How fairylike! We were obliged to go further away. Behind us shone the fires, and they burnt until a strong wind came and extinguished them, which sometimes does not occur for a fortnight or even a month.

'At the extremity of Cape Apscheron is an island called Sviatoï (Holy Island), because, like Bakou, it has naphtha springs. Round the town, wells have been dug, to a depth of from sixteen to sixty-five feet, through a clayey marl, saturated with naphtha. The greater number produce black naphtha, fifteen produce white. Fire is never taken near the springs which are worked, as they would ignite, and it would be impossible to extinguish them. I saw an immense one that was lighted by accident at the beginning of this century; it still burns.'*

* *Voyage au littoral de la mer Caspienne.* By M. Moynet ('Tour du Monde,' No. 8).

In the neighbourhood of Atesh-Gah, the inflammable gas, which is obtained by means of reeds forced into the ground, serves not only for domestic uses, but also to heat the lime-kilns and to burn bodies.

M. A. Ducas, who has given in the *Journal des Mines* some interesting details on these natural sources, quotes an extraordinary phenomenon observed in the vicinity of Bakou: 'After the warm rains of autumn, in the burning evenings, all the country seems in flames. Often the fire rolls along the mountains in enormous masses; sometimes it remains immovable. But this fire does not burn; the traveller taken in the midst of this general surrounding does not feel any heat. Harvests, hay, the reeds remain intact. It has been observed that, during these fantastic conflagrations, the empty tube of the barometer appears on fire, which leads one to believe that this phenomenon is partly electrical.'

SPRINGS OF GAS.

In America, in the state of New York, an infinite number of springs of gas are partly used for lighting. But it is especially in China that human industry has put this phenomenon to profit, and by an ingenious process of working, even obtained both pure water, salt water, and gas to burn, which they conduct to a distance by bamboo pipes, and which they make use of to make salt, warm the houses, and light the streets.

WELLS OF BITUMEN.

Bituminous substances are combustible, liquid, or viscous, sometimes solid, which are divided into several principal varieties—naphtha, petroleum, asphalte, &c. The origin of bitumens has been long discussed, but geologists now agree to regard them as volcanic products.

Petroleums, or rock-oils, nearly always accompany the salses and the giving off of gas which escapes from the interior of the earth. These oils are found in a number of localities—India, France, England, Italy, Sicily, &c. Near the Cape de Verde Isles, petroleum is seen to float on the surface of the sea. In the Gulf of Mexico, the bottom of which has often been shaken by volcanic phenomena, sea-pitch is seen to float in the form of a blackish oil, which shows the existence of submarine bituminous springs. The Gulf of Cariaco, near Cumana, presents the remarkable phenomenon of a naphtha spring which rises above the level of the sea and colours it yellow, over a distance of 1,000 feet. More to the east is the region which, during the great earthquakes of 1766, ejected asphalte and petroleum.

The most abundant springs of petroleum have been discovered in America during the last few years. In 1827, when soundings were being made to find salt springs near Burksville, those engaged saw springing up from a depth of 200 feet an oil-spring, the appearance of which was accompanied with a subterranean noise resembling thunder. The jet rose more than twelve feet above the ground, and the wells being near the Cumberland river, the oil covered the surface to a great distance. A torch being placed near to see if this oil were inflammable, the river was soon seen covered with great flames, which caused serious damage to the proprietors of shore property.

In Pennsylvania there are numerous wells or mineral oil springs. For a long time it has been remarked, in the west part of the State, that the oily matters appeared, from time to time, on the surface of the ground. 'In the course of the year 1859, a farmer named Drake undertook the boring of a well. The farm of Mr. Drake is situated on the banks of a river twenty-eight miles from Meadville. When, in boring, he had reached a depth of sixty-nine feet, he found, instead of water which he looked for, oil in abundance. This was collected by the help of a pump, and, on examining it, it was found to be of very good quality. Other wells were dug for by the side of the first, and gave the same results. The curious flocked by hundreds, and public attention was excited by different accounts. There they saw oil springing out of a hole cut in the rock. A well called "The Chase Well," from the name of its proprietor, exhibited at times real eruptions of oil.

'This discovery has completely transformed the peaceable banks of the river Oil-Creek, which traversed a very picturesque primitive, but scarcely inhabited, country. A few months have sufficed to change all, for Drake's well was only opened in August 1859, and the most important only date from

1860. A crowd of adventurers attacked this land of promise, and have commenced workings on all sides. One might believe oneself in the midst of the camps at California; everywhere are seen carpenters occupied in constructing huts, cart-sheds, and barns, which are not long in becoming a flourishing town.

'The depth at which oil is met with varies from

Fig. 57.—Wells of Bitumen.

30 to 400 feet; the mean is 150. The number of wells now open is nearly 2,000. The oil, as it is collected, is conducted by means of rudely-made canals to immense tubs, which are carefully placed at some distance from the well. This precaution is indispensable, on account of the excessive inflammability of the oil. Frequent conflagrations and accidents have given rise to a law of prudence.'*

* *Moniteur universel.*

THERMAL SPRINGS.—GEYSERS OF NEW ZEALAND.

The formation of thermal springs must be also attributed, in a great part, to the slow, continuous, and regular phenomena of infiltrations. These springs are spread over all the surface of the globe, and are observed from the bed of the sea to the highest strata of mountains. The boiling springs are only found in the vicinity of active volcanoes. We have already described the intermittent springs of Iceland —the geysers; similar springs have been discovered in California, on the eastern slope of the chain of Sierra-Nevada, not far from the lake of Washo. The water rises in jets to a height of 7 yards; the jets follow each other at intervals of five minutes, and produce a noise which resembles thunder.

A learned naturalist, M. Ferdinand de Hochstetter, has recently described, in his *Voyage à la Nouvelle-Zélande,* the numerous volcanoes, thermal waters, and geysers which present many curious aspects and great contrasts in that magnificent country:—

'On the eastern side of lake Taupo are warm springs, to which the inhabitants give the name Waiarikies. I went to the left bank of the Waikato.

'On the two sides of the river, the bushes of the banks cover boiling mud, which must be approached with extreme caution, for the softened ground gives way with the least weight. The largest of these muddy basins has an elliptical form and measures fourteen feet in length, eight broad, and as much in depth. Here a quantity of mud composed of oxide of iron of a bright red was boiling, and viscous bubbles broke out spreading a fetid odour of sulphur. It was a truly infernal spectacle.

'On the opposite bank is found the thermal spring of Tuhi-Tarata. The water, of an azure blue, forms a cascade surrounded with vapour on the steps of tufa, the rows of which descend to the river and which shine with the greatest variety of colours. The same spectacle is reproduced at different points, accompanied by periodical jets at intervals more or less long.

'I went afterwards towards the Rorotua, a volcanic lake fed by thermal springs. In the vicinity of this lake is a small basin, the Rotomahana; it is a real crater of explosion, deep at its centre, edged with swamps to the north and south, a framework of rocks to the east as to the west. It has been rightly named a thermal lake; the quantity of boiling water which flows from the neighbouring springs is so considerable that the entire lake is warmed by them.

'To the north-east is the Te-Ta-Rata, a boiling spring, which, descending from terrace to terrace to the lake, is the greatest wonder of this wonderful country. On the side of a hill covered with ferns at about eighty feet from Rotomahana is the principal basin, the red clayey walls of which are thirty to forty feet in height. It is eighty feet long, sixty broad, and filled to the edge with perfectly pure and limpid water, which, owing to the snowy whiteness of the stalactites on its borders, appears of an admirable turquoise blue, sometimes with opal tints. On the edge of the basin I observed a temperature of 84° centigrade; in the middle, whence the water rises to a height of several feet, its heat was that of boiling water. Immense clouds of vapour which reflected the beautiful blue colour of the basin, wheeled above and arrested attention; one always heard the dull sound of the boiling of water. The man who served us as guide told us that sometimes the whole mass of water is suddenly darted forth with immense force, and that then one can see, at thirty or forty feet deep, the empty basin, which is again filled very quickly. If the fact is true, the spring of Te-Ta-Rata is doubtless a geyser of long intermittence. The water has a slightly saltish taste, but not disagreeable. As in the Iceland springs the deposit is a silicious stalactite. In flowing from the basin, this thermal water has formed a system of terraces which, white as if cut in Parian marble, from a *coup d'œil* of which no description, no image can give an idea; one must have climbed these steps of alabaster, and have examined the peculiarities of their structure to understand how wonderful they are.

'The foot of the hill advances very far in the Rotomahana; above commence terraces containing basins, the depth of which corresponds to the height of the steps of these gigantic ladders: many are two or three feet, some four and six. Each of these steps has a little raised border whence hang on the lower side delicate stalactites, and a platform more or less great which contains one or more basins of a beautiful blue. These form so many natural baths which the most refined art would not have been able to render more commodious or more elegant.

'The highest terrace surrounds a large platform in which are cut many pretty basins of five to six feet deep. In the middle of this platform rises, quite near the principal basin, a rock about twelve feet high, covered with bushes of manukas, lycopods, mosses,

Fig. 58.—The Te-Ta-Rata (New Zealand).

and ferns; it can be climbed without danger, and from it one looks into the blue water covered with vapours from the central basin.

'Such is the celebrated spring of the Te-Ta-Rata. The pure white of the stalactites, which bring out the deep blue of the water, the verdure of the surrounding vegetation, the bright red of the naked walls of the aquatic crater, and lastly its vapour-clouds, which roll over and over and constantly renew themselves, all contribute to form a unique and wonderful picture!'*

INFLUENCE OF THERMAL SPRINGS.

Humboldt was the first to notice the remarkable phenomenon of a thermal spring giving birth to a river charged with sulphuric acid, the Rio Vinagre, which rises at a height of about 10,000 feet on the north-west side of the volcano at the foot of which is built the town of Popayan. The river Vinagre forms three picturesque cascades, one of which falls vertically from a height of 300 feet. There exists also at Java a crater-lake, the waters of which are so strongly impregnated with sulphuric acid that no fish can live there.

Thermal springs are nearly always charged with mineral matters in solution, which they take from the rocks with which they come into contact. They often carry these elements to other rocks, and are thus of great geological importance. M. de Senarmont, in his researches on the formation of valuable minerals, has shown that these springs are not only agents of destruction, but also they transform and create, in circulating in the interior of the earth, and that thus the existence of very important metalliferous veins 'does not always suppose conditions or agents very distant from the actual causes.'†

* *Tour du Monde*, No. 280.

† *Annales de Chimie et de Physique*, t. xxxii. 1851.

By means of the most extended chemical agents in thermal springs, M. de Senarmont has been able to reproduce artificially twenty-nine distinct mineral species, and 'to imitate the phenomena that we see still realised in the places where the mineral creation has concentrated the remains of that activity which

Fig. 59.—Thermal Springs and Fumarolles of New Zealand.

it employed formerly with an energy of another kind.'

We have already spoken of the fumarolles or jets of vapour which are met with on the sides of active volcanoes, in solfataras, and in many other localities. The vapours of which these jets are composed carry also with them different materials which are met with in the water produced by condensation.

Everything indicates that thermal springs and fumarolles, now rather rare, were infinitely more numerous in the old ages of the earth, and contributed to produce the uniformity of temperature, the traces of which we find everywhere. 'Thick fogs must have been spread over the surface of the lands, and hence the radiation into space, such an important cause of cooling, was *nil.* The winters in consequence were less severe, and in this manner we may also explain how many plants and animals, which cannot now endure our climates, could live there then as between the tropics, and precisely like the plants of the south live on the coasts and in the islands of the north, constantly surrounded with thick fogs. All the earth, tempered by these abundant vapours, could everywhere support the same organised beings; and this is why the mineral strata of a certain age present much less difference in the organic *débris* which they contain, in whatever place they are found, than that which exists between the organised beings of different zones.' *

* Beudant, *Cours de Géologie.*

X

UPHEAVALS.

METHANA — SANTORIN — THE ISLE OF JULIA — MONTE-NUOVO — MOVEMENTS OF THE COASTS OF CHILI AND OF THE DELTA OF THE INDUS—SLOW UPHEAVAL OF SWEDEN—LINES OF DISLOCATION—FORERUNNERS OF THE PHENOMENA.

METHANA.

The upheaval of this mountain on the eastern coast of the Morea, between Trezene and Epidaurus, vividly struck the ancients. Historians, travellers, and poets spoke of it. Ovid puts the following description of this phenomenon into the mouth of Pythagoras:—'There is seen near Trezene an escarped and arid peak. The locality was once a level plain; it is now a hill. Vapours shut up in dark caverns in vain seek an opening; under their pressure the soil swelled up like a bladder filled with air, or a bottle made of goat's skin. The earth thus upheaved preserved the form of a high hill, which time has changed into a hard rock.'

This event, to which a date of 223 years before our era is assigned, appears to have coincided with the earthquake which destroyed Rhodes and Sicyon.

Volcanic explosions followed the upheaval, according to Strabo's account. 'An eruption of flames took place near Trezene; a volcano rose to a height of seven stadia. In the daytime it was inaccessible on account of its great heat and sulphurous smell, but at night a more pleasant odour was given off. The heat was such that the sea boiled over an area of five stadia, and twenty stadia away it was agitated and encumbered with blocks of rock ejected by the volcano.'

An error may be granted in the height given to the mountain; as to the agreeable odour which diffused itself round the igneous focus, it has been observed in other volcanic eruptions, and there is reason to believe that it proceeds generally from the presence of naphtha.

Geologists who have visited the promontory of Methone (now called Methana) have established its volcanic formation by the numerous veins of trachyte which traverse it. It has kept the conical form, and possesses also two warm and sulphurous springs.

In the strait which extends between Methana and the calcareous island of Calauria is found a little conical island, in form rather similar to an egg cut in two lengthways; and essentially of volcanic origin. The grey-yellowish trachyte of which it is composed is mixed with lava and scoriæ. It has served for the construction of the modern town of Poros and its port, which was the principal maritime arsenal during the war of Greek independence.

This site, formed by subterraneous fires, is at the present time very beautiful. From it a great part of the Gulf of Athens is seen. It is watered by abundant springs, and the hills are covered with a magnificent wood of lemon trees, in the midst of which rise the ruins of a temple of Diana.

SANTORIN.

The group of Santorin, Therasia, and Aspronisi, which produces the best wines of the Archipelago, is also the locality in which volcanic activity has persisted the longest in that region. At the present time even Nature can there be seen at work.

According to the ancients, these three islands appeared above the waters, after violent earthquakes, several centuries before our era. They form a kind of ring, in which enclosure other islands have risen up at different periods: first Hieras, 186 B.C., then Micra-Kammeni in 1573, and Nea-Kammeni from 1707 to 1712. The following *résumé*, from a narrative written at the time of that last eruption by eye-witnesses, and quoted by Arago, will give an idea of the manner in which these new creations are produced:—

'The 18th and 22nd of May 1707, slight shocks of earthquakes were felt at Santorin.

'The 23rd, at sunrise, there was perceived between two islands, called the Great and Little Kammeni, a body, which was taken for a shipwrecked vessel. Some sailors repaired to the place, and to the astonishment of all the population, a rock had risen

Fig. 60.—THE NEW VOLCANO OF SANTORIN, IN THE GREEK ARCHIPELAGO (1866).

from the waters. In this part the sea was 420 to 520 feet deep.

'On the 24th, many persons visited the new island, disembarked and collected from its surface large oysters, which still adhered to the rock. The island rose visibly.

'From May 23 to June 13 or 14 the island gradually increased in extent and elevation, without shocks and without noise. The 13th of June it was about 1090 yards round, and 22 to 26 feet high. No flame or smoke came from it.

'From the moment of the rising of the island the water was disturbed at its shores; the 15th of June it became nearly boiling.

'On the 16th, seventeen or eighteen black rocks came out of the sea between the new island and the Little Kammeni.

'On the 17th, these rocks had considerably increased in height.

'On the 18th, it sent up smoke, and for the first time great subterranean roarings were heard.

'On the 19th, all the black rocks united and formed one island, but totally distinct from the first. Flames, columns of ashes, and incandescent stones were sent out. These phenomena lasted till May 23, 1708. The black island, one year after its appearance, was five miles round, 2,020 yards broad, and more than 196 feet high.'*

In the year 1866 there was a fresh upheaval in the large crater of Santorin. This appearance was

* *Popular Astronomy.* Arago.

accompanied with earthquakes in the Morea, and possibly they may have been connected with that which was felt (in the middle of May) in the south of France. The particulars which we quote have been given by the learned explorer of Etna, M. Fouqué, sent to examine the region by the Academy of Sciences.

On January 30, dull sounds and slow movements of the soil at the southern extremity of Nea-Kammeni were the first indications of the next eruption. The next day the noises redoubled in intensity, and in the port of the island, called Voulcano, there arose from the sea an innumerable multitude of gas-bubbles.

On February 1, at five o'clock in the morning, the earth divided deeply on the side and at the summit of the central cone. Flames appeared there, as well as at the surface of the sea, which took a reddish colour. The sinking of the earth, on the eastern shore of the port of Voulcano, became very marked on the 2nd. It was necessary to use a boat to enter houses that were before from six to nine feet above the level of the sea. Soon appeared in the port itself, in the midst of a thick smoke, a rock, which during the following days was rapidly changed, but without violent phenomena, into an island, to which was given the name of Georges. The 5th of February it formed a hillock 230 feet long, 100 wide, and 60 high. The blocks which covered its surface were continually thrown out from the centre towards the circumference, as if the increase of the island was from the centre. These blocks were at first black and cold, but they were replaced by others, the temperature of which was higher and higher. Soon they became incandescent, and the entire island appeared luminous in the darkness, with a crown of reddish flames.

All these phenomena increased until the 13th, the time when the land emerged, which was then reunited to Nea-Kammeni, and there formed a new promontory. It not only filled the harbour of Voulcano, but extended into the sea about 200 feet.

It then became the seat of violent detonations, accompanied with projections of incandescent stones. At the same time, at about 164 feet from the coast to the west of the Cape of Phlego, arose another island, which was called Aphroëssa. Its development was slower, and, above all, more irregular than that of the first island. It sank and reappeared alternately three or four times, before taking up a firm position.

During three weeks, abundant projections, accompanied with very strong detonations, took place once or twice every day. George's island, which had attained the height of 164 feet, ejected blocks containing several cubic yards to a considerable distance. One of them caused an accident that spread terror in the island. It set fire to a merchant vessel, after having mortally wounded the captain. On March 10, the eruption had considerably diminished, when a new island, Reka, appeared near Aphroëssa. It did not remain long isolated, as the separating canal, 30 feet deep, was entirely filled up by the 13th.

M. Fouqué's observations permit us to give a very

clear idea of the mode of growth of the volcanic hills of new formation. 'This growth,' says he, 'is certainly partly caused by reason of a slow upheaval of the soil; there are even times when the upheaving action appears to predominate, but it is not ordinarily the case. That which especially contributed to the enlargement of the Georges, Aphroëssa and Reka, are the streams of lava which have their rise there. These streams flow down from either side of the fissure, Georges and Aphroëssa being the two principal points. They advance extremely slowly, cooled as they are on the outside, by the contact of the sea, but nevertheless they flow under the water, which they warm to a temperature near the boiling point. At their surface they present a regular slope on each side of the opening from which they spring, in such a way as very well to represent the two opposite slopes of a roof, of which the line at the top would correspond to the opening. As these streams advance, their thickness increases at any given point of their course, whence it follows that they emerge by degrees, and as their surface is covered with irregular blocks, these appear above the water one after another, and form like rocks round the points which have previously emerged. When, on the contrary, the upheaval of the soil is the principal cause, the blocks which emerge from the water are situated at a certain distance from the centre of activity, and moreover they are always of a low temperature at the time of their appearance, as if the matter which composes them had been for some time solidified. It was in this way that Reka appeared, distant from Aphroëssa some 32 feet, and without the water of the sea being much warmed in its vicinity. At the present time (March 25), there are formed in this manner, by means of upheavals, fresh rocks to the west of Reka, on the side towards Palœa-Kammeni; but, in fact, Georges, Aphroëssa, and Reka, increase principally from the other cause that we have mentioned.'

THE ISLE OF JULIA.

On July 8, 1831, a new island was noticed by the Neapolitan captain Jean Carrao, on a spot between Sciacca, on the coast of Sicily, the isles of Pantalaria, and Malta. It appeared during the shocks of a volcanic eruption. Prince Pignatelli, who observed it on the 10th and 11th of July, remarked that the tall column which rose from the centre illumined the night with a continuous and very bright light. He compared it to a bouquet of fireworks.

The State brig, La Flèche, commanded by Captain Lapierre, was sent to the place by the Minister of Marine. He had on board an eminent naturalist, M. Constant Prévost, who has written an interesting account of this island, which was named Julia; we will give the most important passages.

'On September 25, we reached the island of Maretimo, at the western extremity of Sicily, and at five o'clock in the evening the sailor on watch announced land, from which he saw smoke rise.

'Mounting the tops, we perceived the isle very distinctly, which had very much the form of two peaks united by low land.

'We were eighteen miles distant, and we saw occasionally puffs of white vapour which rose to double the height of the island. Several times, and

Fig. 61.—Crater of the Isle of Julia, Sicily. 1831.

when we were under the wind, we detected a sulphurous smell.

'On the 26th, the wind being contrary and the sea rough, we were obliged to bear off. During the night from the 26th to the 27th, we were overtaken by a frightful tempest: the waves washed over the bridge, and there was no part of the horizon which was not lighted up with electricity and lightning; the thunder rolled continually, but without loud reports. I nevertheless passed the night with my eyes fixed on the volcano, to see if any light proceeded from it; but I did not perceive any indication of a luminous eruption; only the nearly suffocating sulphurous smell which now and then reached the vessel.

'On the morning of the 27th, we succeeded in getting nearer, in spite of an agitated sea. The island, which we sailed round, appeared like a solid black mass, having here the form of an elliptical dome, with a base equal to three times its height, there that of two unequal hills, separated by a large valley: the sides were peak-shaped, except where the vapour escaped in abundance either from a cavity very near the sea, or from the sea itself to a distance of nearly 40 feet. The yellowish-green of the water, modified by the subterranean volcanic action, contrasted with the indigo blue of the open sea, and indicated near the island either rapid currents or reefs.

'On the 28th, the sea having calmed a little, the captain placed a boat at our service. We placed it under the command of M. Fourichon, his second officer, a *Lieutenant de frégate* (now *Vice-Admiral*), and of M. de Proulereoy, first-class cadet. I embarked with M. Joinville, an artist, and rowed by eight experienced and courageous sailors, we arrived in less than an hour near the breakers. We saw that these were produced by the waves breaking with force against a shore, terminated suddenly by a rapid slope, and not by solid rocks. The yellowish-green water in which we were, and which was covered by a reddish foam, had a decidedly acid smell, or at any rate less bitter than out at sea; its temperature was also higher, but only a few degrees (from 21° to 23° C.). We sounded at thirty fathoms from the shore, and we found the bottom at from forty to fifty.

'We directed our course towards the only spot where from the surface of the island we could descend by a gentle decline to the sea. The waves rebounded, rising from twelve to fifteen feet, when they struck the shore. Thirty feet on our left, these waves rose into the atmosphere in a state of vapour; at the same distance to the right, the sea seemed to break on a bank which extended more than a mile in width. The sailors all were agreed that it would be imprudent to attempt to disembark at this time, and that we should inevitably be upset.

'We were but forty fathoms from the island; I could, at this distance, be certain at least for the portion that we had before us, that it was formed of separate and pulverulent materials (cinders, scoriæ) which had fallen after having been thrown into the air during the eruptions.

'I did not perceive any indication of upheaved solid rocks, but I distinctly distinguished the existence of a funnel-like crater, nearly central, from which rose thick columns of vapour, of which the walls were plastered with white saline efflorescent matter.

'We were leaving with regret at not being able at least to carry away some specimens of this new and frightful soil, when a sailor proposed to swim to shore. We tied a sounding-line to him, and in a few minutes, after having at first disappeared under the waves, and in the thick vapour which escaped from them, he arrived safe and sound upon the beach; he made us a sign that that was so burning that he could not keep his feet on it.

'M. Fourichon could not resist the idea of going himself to find some specimens; he threw himself in to swim, and was followed by M. de Proulereoy and by a second sailor, who carried with him a basket, a hammer, and a bottle. I regretted extremely not being a swimmer good enough to be able to follow their example. I remained in the boat, and in spite of its tossing, M. Joinville and I made several rough sketches.

'Our intrepid companions reached the border of the crater, walking on the burning cinders and scoriæ, and in the midst of the vapours which were given off from the surface. They told us that this crater was filled with a reddish and boiling water, forming a lake about twenty-four feet in diameter. At last they returned, after having passed the basket of specimens to us by means of a cord.'

In another expedition, undertaken on September 29, the explorers were enabled to land on the island with a boat, and draw it up. The staff officers of La Flèche nearly all went on shore with M. Constant Prévost, and finished exploring, the work being divided among them. Some measured the circumference, which they found to be 760 yards, with a height of 230 feet; others took thermometrical observations, sounded the crater, or sketched. The tricolour flag with an inscription was hoisted on the highest point of Julia.

'All the island,' says the learned geologist, 'seemed to me to be, like all eruptive craters, a conical mass round a cavity also conical, but reversed. Indeed, by examining the inner walls of the crater, it could be seen that they had a slope of about 45°; and from the lateral fissures, produced by the falling down of the sides, it may be seen that the stratification is parallel to this line of slope, whilst, from the outer side the same materials are placed in an inverted order.

'As to the cliff-like form of the shore, it is easily seen that it is due to the later effect of the falling away of the materials, caused either by the shocks or more probably by the action of the tides, which, drawing away the moveable materials accessible to this action, have successively undermined the shores. These being out of the perpendicular have fallen; every day they become disintegrated, and it is by these means that a beach is formed round the island, a kind of band from fifteen to twenty feet in breadth which terminates with a sharp declivity towards the sea. With this explanation it is easy to observe that these slips continuing every day, the island will gradually become lower until a high tide will carry away that which remains above its level, and there will be nothing in its place but a bank of volcanic sand, by so much the more dangerous as it will be difficult to discover at a distance.

'Everything leads me to believe that the volcano has produced streams of submarine lava, and if, as is possible, the appearance of a crater of eruption was preceded by an upheaval of the soil, which appeared to have been from five to six hundred feet below the level of the sea, there ought to exist round the isle of Julia a belt of upheaved rocks, which would be the border of the crater of upheaval. Perhaps this new disposition of the bottom is the principal cause of the particular greenish-yellow colour of the sea at a good distance from the island, and of the currents which appear in its vicinity.'

This prophecy of M. Constant Prévost was soon confirmed. Already, at the end of December 1831, there was nothing in the place of the island but a bank covered with three yards of water. The volcanic matter had been swept by the waves, and that which remains is the rocky bottom of the sea raised by subterranean forces.

Arago* proves the reality of this upheaval, calculating, according to the soundings of Commander Lapierre, the inclination of the submerged portion comprised between the shore and the corresponding point where the soundings stopped. Finding the slopes much greater than those of the upper cones of Etna and Vesuvius, he thought that with such a declivity, cinders and small stones, supposing that the island had been formed of them, could not have protected themselves against the action of the tides during entire months. An English captain, approaching the island a short time after its appearance, had a thermometer immersed in the sea; it fell progressively

* *Popular Astronomy.*

nearly six degrees; which would follow, according to Arago, from the presence, near the surface, of deep rocks, cooled for centuries, and suddenly upheaved.

We have spoken of the upheaval of a very large island near the northern point of Unalaska in the Aleutian Isles. Other phenomena of the same kind, accompanied by circumstances nearly identical to those which we have described, have been manifested in different parts of the globe. Islands have been seen to rise at different times round Iceland; others emerge at periods of from eighty to ninety years near San-Miguel, one of the Azores. The last, Sabrina, dates from January 30, 1811; its appearance was the prelude to terrible earthquakes in America.

Sometimes the upheaval takes place without solid rocks appearing above the water, as happened in Kamtschatka, when after a great ebullition, accompanied by jets of vapour, the existence of a chain of rocks near the surface was detected, where before there was a depth of 200 yards. A similar phenomenon was manifested only by the heat of the water, in 1820, in a bay of the Isle of Banda, which forms part of the Moluccas; this bay, whose depth was 650 feet, was filled up by a gentle upheaval of basaltic matter, which formed a promontory of some considerable height.

MONTE-NUOVO.

Many savants think that this new mountain (Monte-Nuovo) which rose in the month of September

1538, on the borders of the Bay of Baia, and whose appearance was accompanied with terrible volcanic phenomena, encloses a solid nucleus. According to them, this portion of the materials was raised up in a mass and afterwards opened to send forth the cinders and stones which completed the formation of the cone. According to others, the whole Monte-Nuovo is the result of eruptions from the crater.

According to the testimony of an eye-witness, Francisco del Nero, the earth swelled until it formed a hill, and this circumstance recalls the terrible upheaval of the Jorullo on the Mexican plateau. But against this it may be urged that in this case the walls and the columns of the temple of Apollo, which are near the foot of the mountain, could not have remained perfectly vertical, as they have been proved to be. It is also certain that the volcanic explosion was accompanied by an elevation of the whole level of the Bay of Baia. On the steep rocky shore of Pouzzoles, a crevice filled with marine shells, thirty-six feet above the actual level, establishes this fact.

Another proof of this upheaval is also found in the retreat of the sea to some distance from the shore, a fact affirmed by many witnesses, among others by the savant Porzio, whose interesting account we will here give: 'This region was disturbed during nearly two years with violent earthquakes, so much so that there did not remain a single house intact, no edifice which was not threatened with near and inevitable ruin. But the fifth and fourth days before the calends of October, the earth trembled without intermission night and day. The sea retreated for about two hundred yards; on the dry beach the inhabitants took a quantity of fish, and noticed springs of fresh water. At last, on the third day, a great part of the land comprised between the foot of Monte-Barbaro and the sea was seen to upheave and to take the

Fig. 62.—Monte-Nuovo, Bay of Baia (1538).

form of a rising mountain. The same day, at the second hour of the night, this upheaved ground was transformed into a crater, and sent forth with great convulsions torrents of fire, scoriæ, stones, and cinders.'

'The stones and cinders,' says another account, 'were ejected with a noise similar to discharges of large artillery, in quantities which promised to cover the whole globe, and in four days there was formed in the valley situated between Monte-Barbaro and the lake of Avernus a mountain of at least three miles in circumference, nearly as high as Barbaro itself; to those who did not see it, the formation of such a mountain in so short a time, was a thing almost inconceivable.'

In his observations on the same phenomenon Jacobeo de Tolède adds: 'Some of the stones were as large as an ox; the largest were thrown into the air as far as the range of an arquebuse above the opening, then falling again, some on the border, others into the interior of the crater. The mud thrown out, formed of cinders mixed with water, was at first very liquid, then less so, and so abundant that, with the stones previously mentioned, a mountain of 1,000 yards high was formed by the third day. I climbed to the summit, and looked at the bottom, in which the stones seemed to boil like water in a large caldron on a fire.'

MOVEMENTS OF THE COAST OF CHILI AND OF THE DELTA OF THE INDUS.

The frightful earthquakes that took place in Chili in 1822, 1835, and 1837, destroyed several towns, among others Valparaiso, Melpilla, Quillota, and Casablanca. At the same time many parts of the coast, comprising an extent of more than 200 leagues, rose above the surface of the ocean.

On a shore where the tide never rises more than three to six feet, every movement of the soil can be easily proved. Near Valparasio, at the mouth of the

Concon, to the north of Quintero, some rocks, formerly constantly covered with water, rose six feet above the sea level. On visiting them, oysters, muscles, and other shells were found, the animals themselves being in a state of putrefaction. It was proved that the entire banks of the lake of Quintero, which communicates with the sea, rose more than three feet. Many well-known anchorages diminished in depth. A vessel that was wrecked on the coast, and the remains of which were visited by boat, was found entirely dry after the earthquake of 1822.

Similar occurrences were observed in 1819 in the Delta of the Indus, whilst the country was agitated by violent shocks. Around the fort of Sindree an extent of country has sunk down larger than the lake of Geneva. The village and fort, however, remained standing, and the next day the garrison crossed the sea in boats. Whilst this depression was taking place, a hill fifty miles long and sixteen miles broad was formed in a plain situated to the north. The inhabitants have called it Ullah Bund, or God's Rising. In many places the eastern mouth of the river became deeper. The river, at first turned out of its course, left its bed in 1826 and cut a more direct passage dividing the Ullah Bund.

SLOW UPHEAVAL OF SWEDEN.

It cannot be denied that at the present time the soil of France, except some passing shocks of earthquakes, is in a state of perfect immobility; but the last movements which have completed the elevation of this country above the ocean and have given it its present extent, go back to an epoch which, although doubtless anterior to the historic ages, is not, however, so remote that it is lost in the night of time. The country of Touraine and some of the provinces in the south are covered with a stratum similar to an ocean bed, and on their surface are shells like those which still live on the shores. In the great plains of Picardy, formerly filled with large lakes and swamps, the bones of beavers are found which formerly constructed their dwellings there; and in the bottom of the peat-holes are sometimes discovered canoes cut from a single block, like the ones used by the savages of America, which shows what was then the degree of civilisation of the inhabitants of these parts now elevated above the sea and fertilised by fine cultivation.

'But if we are immovable, and if our frontiers will no longer carry on these conquests and peaceful invasions on the empire of the sea, we have near us countries that do not imitate us, and give us the example of that which we formerly did. The soil of Sweden and Norway continually rises by an imperceptible movement above the waters of the Baltic Sea. This is an established fact, and to form the best idea of it we must, in imagination, take the bottom of the Baltic Sea by its most northern part, at the top of the Gulf of Bothnia, and with a strong arm lift it in such a manner as to cause the waters to flow towards Denmark, whence they would pour into the North Sea, passing through the Sound and the two Belts. As is easy to imagine, this natural process is exceedingly slow, and it will take still a long time before the Baltic is entirely empty; but still the movement is going on every hour and every minute, and in one hundred years the Baltic Sea will not be what it now is at the present time; it is not what it was in the time of the Romans, who called it, doubtless with truth, a great sea.

'The truth of this singular phenomenon, which is almost incredible, is established as follows:—

'First, at a good distance from the coasts and at a considerable height, shells are found which are still very fresh and well preserved, and are of the same species as those which are still taken on the sea-shore. This is a proof dealing with high antiquity. Now for historic times. There exist songs of the ancient bards which celebrate the exploits of warriors when they went to fish, and which contain the names of the rocks on which they usually fished the sleeping sea-calves; these rocks, on which the sea-calves sported, are slightly raised tables above the water, on which these animals easily climbed and stretched themselves in the sun. Now, the rocks spoken of by the bards, and the names of which are still preserved in the country, are, at the present time, at such a height above the water, that the cliffs which surround them completely remove the possibility of a sea-calf climbing up; they have, therefore, been raised up since the time when the ancient Scandinavians sailed round them to send their arrows at the marine animals which lived there. As to our own era, the thing is still more clear and evident, if it be possible. Marks have been made at high water-mark at the foot of various rocks, in order to serve as guides; and on visiting these marks from year to year, they are found successively raised above the level of the sea. It is not the level of the sea which gets lower, for it would necessarily get lower everywhere in the same way on the coasts of Germany and Denmark, as well as on those of Sweden; but this does not take place. It is, therefore, the bottom of the sea that is raised. At the extremity of the Gulf of Bothnia the total rise of the surface in a century is about four feet and a third; at the lower end of the Baltic Sea, below Stockholm, it is scarcely more than a foot; and, lastly, in the more southern provinces opposite to Denmark the movement is imperceptible, and probably no longer continues.

'From this example it must be seen that to form an idea of the changes which occurred in former ages, when man did not exist on the earth, it is not always necessary to have recourse to odd theories and fantastic hypotheses. It often suffices to consider how Nature works at the present time, with perhaps different effects but, in the main, by similar causes. Nature does not change its processes; it is contented, for new works, to modify them. To explain many phenomena in a simple and true manner, it is sufficient to understand that the form of the earth, already far from a perfect spheroid, still changes in many points, and takes other curves; hence volcanoes, chains of mountains, upheavals, and ancient and modern enlargements of continents and islands.'*

LINES OF DISLOCATION—FORERUNNERS OF THE PHENOMENA.

Upheaval still goes on in other countries besides Sweden—in Greenland for instance. In Oceania vast extents of submarine land, which are indicated by archipelagoes, rise and sink imperceptibly. These phenomena concur with sudden and violent eruptions to modify the form of the earth's surface. We find here, however, but a feeble indication of the mighty work by which our dwelling has been prepared.

Geology teaches us that the systems of mountains have been upheaved at different epochs corresponding to the phases of cooling of the internal part of our globe remaining in the original state of incandescence. The hypothesis generally admitted gives us above the fluid sphere at first a thin coat, gradually thickened by the crystallisation of the rocks on the lower side, and which has received on its upper side, by the successive condensation of atmospheric agents, the waters of the oceans, the elements of the soil, and living organisms.

'The opinion which ascribes to mountains a volcanic origin,' says A. Bertrand,† 'was sure to have been considered very problematical at the time it

* Jean Reynaud. *Magasin pittoresque*, t. i.

† *Lettres sur les révolutions du globe.* 6th edition.

was first enunciated, and those who brought it forward could not have been able to produce the facts necessary to prove it well. Besides, the expression was inexact; it would have been more just if they had stated that the relief of the mountains is in a great measure due to volcanic phenomena, taking the word volcano in the broad sense that M. de Humboldt gives to it. This savant, indeed, defined volcanicity to be "the influence that the interior of a planet exercises on its exterior envelope in the different stages of its cooling," and most geologists at the present time adopt this definition, which does not permit the separation of results one from the other which are due to an identical cause, acting with different degrees of intensity. The first volcanoes of the earth were nearly all opened in the primitive rocks, before the secondary rocks were formed; they have since been covered by the latter, of which the successive formation is evidently due to the sea, or to immense lakes of fresh water. This large quantity of volcanoes belonging to the primitive formations when the solid crust of the earth was thinner is favourable to the opinion which we have just stated. Later, by the double reason of the diminution of activity of the interior focus and the increasing thickness of the stratum that covers it, the eruption of the volcanoes ought to have been much less frequent, and this is indeed the case.'

The terrestrial crust, at present solidified to a depth of about thirty-one miles, forms a screen such that the radiation of the central heat is nearly imperceptible on the surface, but this screen is comparatively very thin—thinner, say, than the shell of an egg—if we compare it to its contents. It is a flexible envelope, as not only the slow movements which are observed prove, but again the numerous folds in most of the mineral strata. Lying directly on the liquid nucleus, and altering its form in proportion as the globe cools, it undergoes a reaction from the nucleus, which endeavours to take the spherical form, by virtue of the laws of attraction. When the forces thus put into play increase beyond certain limits the crust gives way and there is a change on the surface of the globe which, without changing in extent, should correspond to a volume less than that which it enveloped in the first instance. The fractures produced, then, on the lines of least resistance (arcs of circles in the case of a sphere) give rise to the formation of chains of mountains, and occasion numerous changes in the distribution of the continents and seas, and consequently in all conditions of life. A period of repose afterwards prepares fresh revolutions and upheavals.

A great number of clefts, produced in the earth's crust by this series of movements, have been filled at different times by a variety of substances emanated from the depths. It is in studying the arrangement of metallic deposits in the bosom of the earth that miners have been led to the simple law on which are founded the most important discoveries of geology. In each district that they explore veins of the same composition and of the same age are always found parallel. Ought not this similarity of direction, extended to a whole class of facts observed in the terrestrial strata, to apply also to chains of mountains? Such is the question which an illustrious geologist, M. Élie de Beaumont has succeeded in resolving affirmatively after long researches.

Simple relations observed in the angles that the lines thus characterised enclose between them induced the same savant to seek if the whole of the systems of mountains could not be comprised into a regular network spread over the globe. We have explained the function of the hexagon in the crystallisation of the basaltic prisms: it is the pentagon which possesses analogous properties relatively to the sphere.

Not only are the systems of mountains represented by lines of pentagonal network, but these lines also furnish the most useful information on the constitution of the terrestrial crust. They reveal the mineral bearings in unknown regions, the sources of naphtha and petroleum, which have now acquired so much importance, and trace the direction in which soundings should be made to discover them in the depths of the earth. Following, for example, the arc of the great circle which, starting from the salt lake of Seistan, passes near Bakou and Iceland—places remarkable for their bituminous emanations—it will then be seen to lead to the abundant sources of petroleum of Mecca and of Oil Creek in Northern America.

By the development of the important labours

which we can but allude to, we shall learn to know better and better the relative stability of the terrestrial regions, and perhaps the foretelling of volcanic phenomena will become a scientific fact.

We read in the *Bulletin de l'Association scientifique*:* 'The regularity of chemical phenomena which are produced in volcanoes, noticed for the first time by M. Sainte-Claire Deville, explains a crowd of facts which without it would be veritable enigmas. It also gives valuable indications on the periodicity, and even on the probable intensity, of future eruptions.'

The influence of these beautiful discoveries is not only visible in the progress of our security, and of the well-being which results from it, it appears also in the development of the ideas connected with the notion of universal order, and with the Divine action which this admirable order affirms and unveils to us. 'The knowledge of laws,' says Humboldt, 'increases the sentiment of the calm of nature. One would say that the discord of the elements—this great bugbear of the human mind—is quieted in proportion as science spreads its influence.'

We can understand that during the age of ignorance, when the disasters produced by the powerful subterraneous forces struck only the imagination, earthquakes would be attributed to Divine anger, and craters regarded as the breathing-holes of the infernal regions. In the present day we come back to the original idea of the Greeks, that in the awful

* No. 7. July–August, 1865.

mystery of volcanoes is seen Work. This religious presentiment is become Science. By means of it we admire the incessant work of nature; and we regret not to be able to enlarge, as much as we should have wished, on all that is unveiled to us of this fertile and magnificent activity. But in this sketch, in which, in order to induce study, our special duty has been to reproduce the most curious observations and most interesting descriptions of naturalists and travellers, it is only possible to glance at so vast a subject. In trying, however, to afford a glimpse of the grandeur of the laws, the beauty and the general character of the phenomena, their creative influence, the prodigious part played by volcanoes in the formation of the earth's crust, and the actual utilisation of their products, we hope to have contributed to spread a truth which becomes more striking every day; to prove anew, starting from the glorious discoveries of modern genius, that nature is 'that which perpetually grows and develops that which has no life except by continual change of form and interior movement.'*

* Carus.

Printed in the United States
By Bookmasters